COLLECTIVE WISDOM

COLLECTIVE WISDOM

CO-CREATING MEDIA FOR EQUITY AND JUSTICE

KATERINA CIZEK AND WILLIAM URICCHIO

WITH COAUTHORS JUANITA ANDERSON,
MARIA AGUI CARTER, DETROIT NARRATIVE AGENCY,
THOMAS ALLEN HARRIS, MAORI KARMAEL HOLMES,
RICHARD LACHMAN, LOUIS MASSIAH, CARA MERTES,
SARA RAFSKY, MICHÈLE STEPHENSON,
AMELIA WINGER-BEARSKIN, AND SARAH WOLOZIN

THE MIT PRESS   CAMBRIDGE, MASSACHUSETTS   LONDON, ENGLAND

The MIT Press would like to thank the anonymous peer reviewers who provided comments on drafts of this book. The generous work of academic experts is essential for establishing the authority and quality of our publications. We acknowledge with gratitude the contributions of these otherwise uncredited readers.

This book was set in Stone Serif by Westchester Publishing Services.
Printed and bound in the United States of America.

Library of Congress Cataloging-in-Publication Data

Names: Cizek, Katerina, author. | Uricchio, William, author.
Title: Collective wisdom : co-creating media for equity and justice /
   William Uricchio and Katerina Cizek [and others].
Description: Cambridge : The MIT Press, 2022. | Includes bibliographical
   references and index.
Identifiers: LCCN 2021057615 | ISBN 9780262543774 (hardcover)
Subjects: LCSH: Mass media—Authorship. | Creation (Literary, artistic, etc.) |
   Artistic collaboration. | Social justice.
Classification: LCC P96.A86 C59 2022 | DDC 808.06/6302—dc23/eng/20220303
LC record available at https://lccn.loc.gov/2021057615

10 9 8 7 6 5 4 3 2 1

Authors' Note

One hundred percent of the royalties from this book will be donated to Allied Media Projects based in Detroit, Michigan.

Where do you begin telling someone their world is not the only one?
—Lee Maracle

CONTENTS

INTRODUCTION

Co-creation is everywhere and yet easy to miss. It's how the internet was built. It forms the backbone of social movements, systems of knowledge, scholarship, and worldviews. But in the current operating system, individuals take the credit for—and the profit from—these collective forms of authorship, erasing co-creation and whole cultures from the narratives people use to make sense of the world.

Co-creation seems invisible, yet it's nothing new. Take the art of petroglyphs (rock carvings). There's probably a site near you. These sites exist almost everywhere in the world, and yet they are so commonplace, so subtle, that often people don't see them. Etched into stone, cliffs, and rock faces on every continent except Antarctica, they have defied millennia of weathering, surviving as evidence of early collective stories and cultural knowledge, left as traces along the paths of human migration since the dawn of humanity in Africa.

One such prehistoric petroglyph site lies cached on the barren flats of Gobustan, at the edge of the Caspian Sea. At first glance, the site is not easy to spot. It is hidden inside an enormous heap of rocky boulders jutting out of the surrounding semi-desert. Up in the belly of the rocks in crevices and sprawled across its interior rock faces, there's a spectacular collection of more than six thousand prehistoric rock carvings etched over the course of forty thousand years. The carvings were not recorded into modern history until the 1930s, when local workers at a stone quarry stumbled across what is now a UNESCO World Heritage Site.

Collage created by Helios Design Lab, with a photo courtesy of Co-Creation Studio.

Today, archaeologists believe that the carvings suggest a world now difficult to imagine in this location's current dry, dusty, arid climate. They point to a particularly breathtaking collection of carvings of longboats, which may reflect a previous environment abundant with water featuring wetland shores that may have been an ideal habitat for reeds that could sustain this scale of boatbuilding.

Thousands of the other carvings at Gobustan feature human figures dancing, warriors with lances in their hands, antelopes and wild bulls fleeing, battle scenes, caravans of camels, and images of the sun and stars. Here, inscribed in stone, is life on Earth and the cosmos as understood by humanity over millennia. These carvings also provide evidence of the recurrent co-creation processes that have shaped human languages, music, early texts, performance, architecture, and art over the millennia.

Were the carvings made by priests, prophets, and protoprofessional artists or by ordinary people, collectively, over thousands of years? Of the thousands of engravings at the Gobustan petroglyph site, one inscription was likely the work of a single person. It's probably the last carving of note here, a piece of graffiti found at the base of the site. It appears to have been carved by a Roman legionary passing through the region in

Qobustan petroglyph site. Photo by Bruno Girin. Image under Creative Commons.

Qobustan petroglyph site offers evidence of the deep history of co-creation. Photo by Walter Callens. Image under Creative Commons.

the first century CE, a version of the message "I was here." The sentiment feels lonely, almost mournful, when juxtaposed against the collective spirit rising from the petroglyphs across the interior of the massive rock faces and surviving across millennia. What they radiate instead is something joyful and ecstatic, the proclamation "We are here!"

In the past couple of centuries, however, the gazes of Western commerce and scholarship have tended to focus on the singular "I was here." Industrial forms of top-down media privileged the myth of the solitary author, which often served as a rationalization of extractive, harmful, and commodifying practices. By contrast, as Chimamanda Ngozi Adichie said in a 2009 lecture, "When we reject the single story, when we realize that there is never a single story about any place, we regain a kind of paradise."[1] Co-creation is increasingly recognized in such areas as education, health care, technology, and urban design. Although each field has its distinct approaches, fundamentally co-creation is an alternative to—and often a contestation of—the singular voice of authority, control, and ownership.

Perhaps the largest and highest-profile recent act of co-creation started in the unlikely world of biomedical science in 2020–2021 with the rapid response to

COVID-19, the virus that caused the largest global pandemic in more than a generation. Never before had so many scientists around the world joined together to fight one pathogen. This mass collaboration required a radical deviation from the scientific status quo. As Francis Collins, the director of the US National Institutes of Health, the largest funder of biomedical research in the world, told *The Guardian*, "I have never seen anything like this. It has been all hands on deck."[2]

Events went into motion on January 11, 2020, when Shanghai virologist Yong-Zhen Zhang broke a government embargo to release the genetic code of COVID-19 on the website virological.org. Only days earlier, his team had cracked the code in a stunningly short forty hours, using the latest high-throughput sequencing technology for RNA. They discovered a pathogen resembling SARS, and Yong-Zhen immediately sounded public health alarms in China and posted the code to the US National Center for Biotechnology Information. But such formal submissions to the institution can take weeks to process, and Yong-Zhen and his colleagues worried that the world needed to move more quickly. Together, they decided to go public on an open-access discussion forum on the web.[3]

This data drop opened the floodgates to scientific cooperation across national borders, across scientific disciplines, and, significantly, beyond legal intellectual property conventions in search of effective vaccines and treatments for COVID-19. Suddenly, thousands of scientists were forgoing future claims to patents and prizes, putting aside all other research to jump in, share data openly and often in real time, and collaborate with former competitors. The free code to the virus had emboldened and armed them with the knowledge they needed.

The Massachusetts Institute of Technology (MIT), where the authors of this book are based, shuttered all its research labs on March 14, 2020, except for those dealing with COVID-19. Anyone who could help climbed on board. Teams toiling on decades-long research into mRNA vaccines pivoted to go after the coronavirus. Labs that had 3D printing developed and released DIY patterns for ventilators. Mathematicians helped support complex modeling teams to predict the spread of the disease. CRISPR pioneers Jennifer Doudna and Feng Zhang, who had been battling each other in court for years over patents, put aside their disputes to ensure that their innovations in gene-editing technology would contribute to COVID-19 testing methods.[4] Walter Isaacson, historian and author, tracked Doudna and Zhang's relationship in his 2021 book *The Code Breaker*. In the book, Isaacson characterizes the cutthroat pre-crisis situation in the sciences:

Before the pandemic, communication and collaboration between academic researchers had become constrained. Universities created large legal teams dedicated to staking claim to each new discovery, no matter how small, and guarding against any information sharing that might jeopardize a patent application.

"They've turned every interaction scientists have with each other into an intellectual-property transaction," said Berkeley biologist Michael Eisen. "Everything I get from or send to a colleague at another academic institution involves a complex legal agreement whose purpose is not to promote science but to protect the university's ability to profit from hypothetical inventions."[5]

The urgency of COVID-19 changed these rules. However, like any process of co-creation, it was messy and somewhat risky. Inaccurate or poorly conducted research has at times muddied the waters, incomplete and evolving research findings made consistent public health messaging a challenge, and while scientists blew the doors open to co-creation in their labs, other systems further downstream remained tightly controlled, reproducing inequity and injustice on local and global scales. Who on the ground ended up being tested, treated, and vaccinated? The actual production, supply chains, and distribution of COVID-19 tests, treatments, and vaccines have revealed how quickly co-creation can become co-opted and exploited to benefit a few individuals, corporations, or countries, conforming to hierarchies of race, class, and geopolitics. Also proven to be chaotic is having the public witness scientific discovery unfold in real time, as raw complex data is imperfectly interpreted by rushed or compromised journalists and, less transparently, WhatsApp and Facebook groups, among others. Public trust in science was eroded by these events and became even more of a divisive partisan issue than it already was.

Such mis- and disinformation has become its own kind of pandemic in a highly weaponized mediascape. Social media platforms and artificial intelligence tools (such as facial recognition and big data), now broadly understood within the framework of surveillance capitalism,[6] have become particularly insidious because they are cloaked in a false veneer of neutrality, objectivity, and benevolence of technology. *Algorithms can't lie, can they?* Prior to the pandemic, Ruha Benjamin in her book *Race after Technology* (2019) urged society to confront "the New Jim Code," the way in which algorithms and tech design reproduce racial discrimination, in parallel to the Jim Crow laws of racial segregation in the southern United States in the nineteenth and twentieth centuries. Similarly, Safiya Umoja Noble in her 2018 book *Algorithms of Oppression* exposed how search engines reinforce racism. While the scientific community may have met the moment of COVID-19 head-on through co-creation, society has yet to focus its collective efforts to reign in tech-enhanced systems of oppression. We don't have an app for that.

Genomic epidemiology of novel coronavirus - Global subsampling

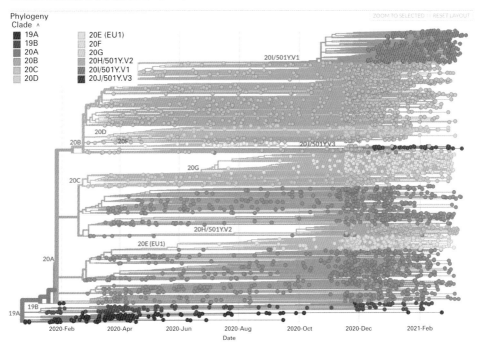

Nextstrain is an open-source project designed to track pathogen genome data. The image above shows 3,872 genomes sampled for SARS-Cov-2 analysis between December 2019 and March 2021. The project is made possible through the sharing of data from research groups around the world, a collective venture arising from scientists and labs working together. Screenshot used under a CC-BY-4.0 license. Courtesy of nextstrain.org.

"Technology can put a man in space, or a nano-computer in every creature on Earth. Yet technology cannot answer this question, that should be asked of anything, and it is an Indigenous question: 'Is it wise?'" said Ojibwe elder, artist, and scholar Dr. Duke Redbird at the Open Documentary Lab's lecture series in 2020. "Wisdom is not a topic that is taught or studied in the curriculum in our schools or universities, nor is it a practice in modern life."[7]

This book is titled *Collective Wisdom* because we're striving to consider the crucial question that Dr. Redbird urges everyone to address: "Is it wise?" In these pages, we'll examine co-creation tools aimed at building the media we need. We gather the voices, works, and learning of hundreds of folks from divergent walks of life who insist that co-creation becomes wise only when it is tied to equity and justice.

Consider the term *collective intelligence* coined by cultural theorist Pierre Levy, which he says is understood as shared group intelligence emerging from the collaborations, collective efforts, and sometimes competition of many individuals, often through processes of consensus decision-making.[8] The concept of collective intelligence has been applied to bacteria and animals, especially hived insects. Recently, *collective intelligence* has been used to characterize crowdsourcing and the potential of computer systems, as explored by MIT's Center for Collective Intelligence. In this book we are interested in collective intelligence because it provides us with a system of tools. However, the term *collective wisdom* goes further, evoking the spiritual and the philosophical, questioning the powerful, and broadening our range of questions about not only *how* to co-create but also primarily *why* and *why now*.

For us, media co-creation is defined by methodologies that offer alternatives to the singular-author vision and seek collaborative routes to discovery and toward justice. The spirit of co-creation allows for projects to emerge from process rather than the other way around. Co-creation accounts for the uncomfortable yet crucial understanding that people don't come to the process on equal terms and seeks to decolonize the systems that oppress and to stimulate projects that don't simply document or passively observe the world but instead insist on changing it.

We intentionally drew the stories of the Gobustan petroglyphs and the COVID-19 vaccine from unexpected contexts: prehistory and institutionalized science. Most of this book, however, focuses on the radical, sustained practices of co-creating media within communities and with social movements. We also extend our definitions to address the urgent need for co-creation across disciplines and organizations as well as in concert with nonhuman systems in biology and technology. We aim to connect all these unusual dots in order to build a field guide for co-creation tools. What emerges, we hope, is the elegance of collective wisdom—a shared, decentralized understanding that, when intentionally channeled, can lead to transformative shifts that ripple out beyond the places where they originated.

Today, though, co-creative practices are still mainly hidden in the margins and are rarely acknowledged, documented, credited, and accounted for. Co-creation projects are buried and overshadowed by prevalent conventions that support the individual author, a comparatively recent phenomenon that is now being challenged by the digital era. It is not that we seek to displace singular authorship in media making altogether. Rather, we are interested in opening up parallel pathways for the funding, institutional support, celebration, distribution, and sharing of these complexes of collective practices that currently exist in the margins.

We Live in an Ocean of Air *(2019). Used with permission from Marshmallow Laser Feast.*

When we use the word *we* as authors in this book, who do we mean? In the interest of accountability, we acknowledge that due to institutional constraints and proximity, this study has two primary authors based at the MIT Co-Creation Studio—one a documentarian and the other a scholar—whose roots are not in the social locations and communities from which a lot of this work emerged. In recognition of that gap, we have tried to design our approach so it is informed by and reflective of a wide variety of other perspectives. Our research team was multidisciplinary, composed of journalists, technologists, place makers, researchers, and students who are concerned with the history, relevance, and opportunity of collective methods. Additionally, we intentionally sought out the expertise of members of historically marginalized communities both within and outside academia. We invited coauthors to write chapters and shape the conversations, and in one case we were invited to enter into a written community-benefit agreement. Finally, we embarked on an extensive participatory reviewing and editing phase of this book. Overall, we conducted ninety-nine individual interviews and held ten group discussions; a total of 166 people working in media and related fields participated. No such process can be comprehensive, but we put a priority on listening to practitioners in the field rather than starting with a theory that we sought to prove. The geographic scope of our research was mainly

The term collective intelligence *is often applied to insects, such as termites, pictured here in their cathedral termite mound. Photo by J. Brew. Used under Creative Commons.*

limited to North America, although several of the projects and people involved reside elsewhere in the world. Further, many participants referenced work tied to their ancestral or diasporic communities. And at our MIT-based Co-Creation Studio, we've sought to incubate projects that would continue to test and deepen our insights.

Many of the questions posed in this study are based on the authors' own twenty years of experience within co-creation. This includes Katerina Cizek's decade-long sojourn at the National Film Board of Canada, where she worked on two long-form, co-creative, cross-disciplinary documentary projects that involved in-person and online communities as well as nonhuman systems. William Uricchio's encounters with co-creation range from working with communities and first-generation Portapaks (early portable video recorders) on documentary production to helping create the "undisciplined," lab-centric, publicly engaged Comparative Media Studies program at MIT.

Most importantly, this work is shaped and coauthored by many people. Among our many co-creators in this work, author and place maker Jay Pitter contributed valuable key questions and framing. The book includes chapters that center the first-person voices of the Detroit Narrative Agency, Amelia Winger-Bearskin, and Louis Massiah as well as an extended excerpt from a conversation among Thomas Allen Harris, Michèle Stephenson, Maori Karmael Holmes, Maria Agui Carter, and Juanita Anderson and quotations from more than one hundred other interviews and discussions. The book also features papers, analysis, and spotlights (case studies) written by Sarah Wolozin, Dr. Richard Lachman, and Sara Rafsky and includes vibrant examples of hundreds of projects such as the landmark co-creation projects discussed below, which suggest the diversity and abundance of approaches to co-creation.

LANDMARK CO-CREATED PROJECTS

What do we mean by co-created media projects? There are ten landmark projects that exemplify the diverse co-creative approaches we describe in *Collective Wisdom.*

Used with permission from Kamal Sinclair.

Question Bridge. This is a documentary project in which the co-creators invited Black men across the United States to record questions for subsequent interviewees as well as answers to previous interviews. The project has taken many forms, including a five-channel video installation, a book, a mobile app, and community events.

Used with permission from Isuma Distribution International.

SGaaway K'uuna (Edge of the Knife). This is a dramatic feature film shot entirely in the Haida language (British Columbia, Canada) and was co-created by three organizations: the Haida Nation governmental body; Isuma Productions, a Canadian Inuit production company; and the University of British Columbia.

Used with permission from the International Consortium of Investigative Journalists.

The Panama Papers. A global collective of investigative journalists from 107 news organizations joined forces in 2016 to interpret the biggest data leak in history, exposing worldwide corruption. The Panama Papers brought down governments and marked the largest-scale effort ever in journalism to collaborate rather than compete.[9]

Used with permission from Thomas Allen Harris.

Family Pictures USA. Artist Thomas Allen Harris co-creates a living and growing American "family album" by traveling across the country and inviting community members to share images and stories from their personal family archives. The resulting work involves live interactive performances, documentary films, web projects, and a 2020 TV special series on the US-based Public Broadcasting Service television network.

Used with permission from Sougwen Chung.

D.O.U.G. Artist Sougwen Chung co-creates paintings with robots in front of live audiences, prompted variously by the artist's actions as well as live data from urban surveillance systems.

Used with permission from Zhang Mengqui.

The Folk Memory Project. Based in Beijing, China, this collective invites young film-makers to document the experiences of their relatives and elders in rural communities during the Great Famine of 1959–1961. This growing body of work now includes over one thousand interviews. The collective creates films and performs the recordings to live audiences, using projection, dance, and multimedia.

Used with permission from James Minton.

Eviction Lab. This is a co-created, transdisciplinary project based at Princeton University that draws on the collective expertise of sociologists, statisticians, economists, journalists, web engineers, and community members to help document the rising crisis of evictions across America in real time.

Used with permission from the National Film Board of Canada.

Challenge for Change. This thirteen-year-long government-sponsored program (1967–1980) at the National Film Board of Canada experimented with a wide array of co-creative methods, including participatory media, to help historically marginalized communities tell their own stories. The program resulted in over two hundred films, including the famous collection of twenty-seven titles made on Fogo Island, and *You Are on Indian Land* (1969), filmed by the first known all-Indigenous film crew.

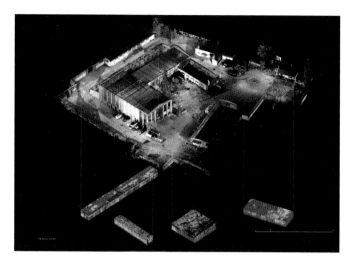

Used with permission from Forensic Architecture.

Forensic Architecture. Since 2010, this co-creative research agency has undertaken more than seventy investigations into human rights violations including violence committed by states, police forces, militaries, and corporations. The investigation teams include architects, software developers, filmmakers, investigative journalists, artists, scientists, and lawyers. They are developing techniques in spatial and architectural analysis, open-source investigation, digital modeling, and immersive technologies as well as documentary research, situated interviews, and academic collaboration.

Used with permission from the National Film Board of Canada.

HIGHRISE. This seven-year project led by Katerina Cizek at the National Film Board of Canada (2008–2015) brought together documentarians with architects, urban planners, place makers, technologists, housing activists, geographers, and residents of high-rise buildings to chronicle the vertical and virtual lives of suburban high-rise communities around the world. The work resulted in digital documentaries, art installations, books, local interventions, and live performances as well as policy changes at local levels.

Books necessarily bear the traces of their time. *Collective Wisdom: Co-Creating Media for Equity and Justice*—the research that it is based on and the co-creative methodology it interrogates—is no different. Rooted primarily in the North American experience, the research and writing unfolded at a traumatic moment marked by renewed struggles for racial justice in the face of police brutality, the COVID-19 pandemic, a mediascape riddled with misinformation and disinformation, ever more extreme manifestations of climate crisis, and even an insurrection in the US Capitol following the electoral loss of a celebrity real estate developer turned president. These events took millions of lives.

The complexity of these issues has exposed the limits of the coping mechanisms of traditional institutions, with their narrowly focused domains of expertise. More fundamentally, society's ineffective responses to each of them reveal the deeper problem of an epistemological rift manifested in debates over what is real and what is not. This breakdown in shared notions of the real and in the basic protocols of everyday life combine to make the work of this book more urgent. Co-creation, we argue, offers ways to repair and rebuild trust and craft shared visions. In addition, co-creation offers approaches to drawing on collective experience and tackling complexity. While co-creation is not a panacea—like any method, it can be put to counterproductive ends—it offers a way to discover what is common in fractured worldviews, helping to restore shared visions.

The pages ahead chart co-creation as a *practice* with deep historical roots and myriad cultural manifestations. The chapters consider co-creation as a resurgent *methodology*, with expressions in fields as diverse as industrial design, social movements, and planning. In addition, the chapters explore co-creation as a *concept* that resonates in the work of such diverse thinkers as Émile Durkheim and James W. Carey on communication as ritual and exchange rather than transmission and extension. Above all, we are interested in pursuing the potentials of co-creation for equity and justice in a world marked by polarization and oppression and paralyzed by contested epistemologies.

Collective Wisdom aims to bring the ongoing work of the MIT Co-Creation Studio to a wider public realm. The book is the assembled result of many years of thought, work, and dialogue helping to propagate new ways of working and it's our version of the message that echoes in the chorus of thousands of rock carvings at Gobustan proclaiming "We are here."

1

"WE ARE HERE": STARTING POINTS IN CO-CREATION

What exactly do we mean when we say "co-creation" in the context of media? Through our research and interactions with practitioners, we have arrived at this definition: "Co-creation offers alternatives to a single-author vision, and involves a constellation of media production methods, frameworks, and feedback systems. In co-creation, projects emerge out of process, and evolve from *within* communities and *with* people, rather than being made *for* or *about* them. Co-creation spans across and beyond disciplines and organizations and can also involve non-human or beyond human systems."[1]

The concept of co-creation reframes the ethics of who creates, how, and why. Our research shows that co-creation interprets the world and seeks to change it primarily through a lens of equity and justice. Our definition is not intended to be prescriptive or to claim territory. Rather, it is an offering for those who may find it useful and an attempt to articulate a shared language and best practices across divergent fields and locations.

THE MAKING OF A TERM

Words bear traces of their historical making, but the word *co-creation*, at least in English, seems to be something of an anomaly. The *Oxford English Dictionary* has not yet incorporated the word. However, the Ngram Viewer, based on Google's digitized book holdings, locates the term's appearance as early as the mid-nineteenth century, in horticulture, with rapidly escalating usage after 1960. There are many criticisms of Ngram, including that it draws on a limited and culturally loaded knowledge base and is distorted by the flaws in the optical character recognition technology it relies on. Still, Ngram gives a sense of how the term has been deployed. Ngram's links include texts on sales organizations ("a marketing concept requires co-creation. It calls for living with your customer on all his decision-making and influencing levels, at all stages of his planning"),[2] philosophy, spiritualism, the family, jurisprudence, and music ("Mr. [Paul] Hindemith calls the process by apt but agonizing Americanisms like 'co-construction' and 'co-creation'").[3]

More recently, though, the term *co-creation* has been picked up by social movement proponents, management specialists, and big-tech entrepreneurs alike. Unsurprisingly, the term enjoys some semantic slippage. *Co-creation* can mean something as general as taking part in a cool new movement, or it can be unpacked into nuanced and competing senses that share certain principles while asserting their own particularities (see the co-creation wheel of associated practices later in this chapter). Obviously, however, the word bears a close relationship to some more familiar terms, such as *collaboration, communal and collective creation, grassroots,* and *participation.*

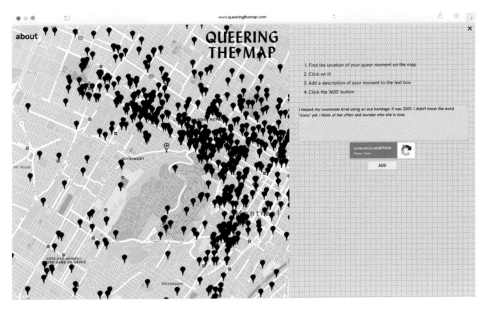

Queering the Map (2018). Used with permission of Lucas LaRochelle.

"Participation, for me, assumes that there's one person who's created the art that people are participating in, but ultimately, it is that person's project," said Lucas LaRochelle, who initiated the crowd-sourced cartographical project Queering the Map. "Co-creation," on the other hand, "would be a more distributed network of actors coming together to create something."

Co-creation overlaps with other terms used in particular fields. *Co-design*, for example, implies the collective build of a tool or a platform rather than the co-creation of a project, according to Amelia Winger-Bearskin, an artist and curator. Coauthorship describes a wide range of practices, only some of which could be considered co-creative. While some coauthors submit their distinct chapters to each other without discussion, others conceive and write every sentence together. *Coproduction* is often used in radio to describe co-creation but in the realm of film refers specifically to contracted obligations between production houses, often in different countries or territories.

What accounts for the fuzzy borders and myriad uses of the word *co-creation*? Has it been excluded from dictionaries because it runs counter to dominant cultural norms or perhaps because the practices it refers to are so intrinsic to human social behavior that they are easily overlooked? Perhaps the rise in its usage is symptomatic of a change in the status quo of cultural production and dramatic shifts of assumptions that require a new term to mark them.

A MYRIAD OF PERSPECTIVES

At its core, co-creation is relational. The word *co-creation* connotes a collective sense of ownership, a joint journey of discovery, and an abandonment of ego. While the participants in our interviews had a wide range of definitions for and connections to the term, they all agreed that co-creation is not for everyone, every project, or even every stage of every project. But discussing its meaning leads to deeper questions about how media works are or can be created.

First of all, when is a media project definitely not co-creation? And how does co-creation differ from participation or collaboration? "If it's about one person's need to tell their particular version of that story, I don't think it's co-creation," said Nell Whitely of Marshmallow Laser Feast, an immersive storytelling studio in London. Nicholas Pilarski, of the Peoples Culture arts collective in New York, reinforced that point: "I think if you're in this to articulate your own personal viewpoint of the world, and if that is your chief desire, then you don't go after co-creation. . . . We're after learning something with somebody, [and] going through a process." Gina

"We Live in an Ocean of Air" (2019). Used with permission from Marshmallow Laser Feast.

Czarnecki, a British pioneer in bio-art—in which creators work with living tissues, bacteria and other organisms, and life processes—specifically distinguished between collaborating with scientists and co-creating with them: "For me, [*co-creation*] means developing the ideas from concept, [the] original concept. I suppose that's pure co-creation. The other form is finding people who can develop and enhance the project into something that you could never have anticipated yourself. It's better than the sum of the parts, and taking both of your disciplines and strengths and co-evolve it into something together."

"Co-creation is at the beginning of the process. Everyone's there from the very beginning," said Toby Coffey of the National Theatre in London, "while collaboration is one person coming into the room with an idea that brings a group of people together."

Elizabeth Miller, documentarian and scholar at Montreal's Concordia University, said that she thinks of these practices "as a continuum. Participation can be consultation, but co-creation feels like the side of participation that really, deeply engages with partnerships." And Babitha George, a designer and festival founder based in Bangalore, proposed that "co-creation is really about allowing your own practice to be

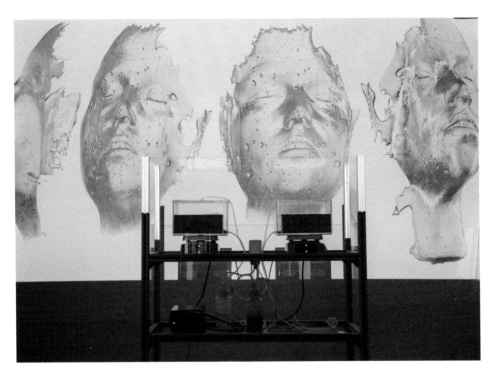

Installation view of Heirloom (2016), Gina Czarnecki and John Hunt. Used with permission from Gina Czarnecki.

A still from The Other Dakar, by Selly Raby Kane, which is part of Electric South, a South African non-profit that supports a network of artists across Africa to tell immersive, interactive stories using virtual reality, augmented reality, and other digital technologies. Used with permission from Electric South.

examined and questioned, based on perspectives from other people and other practices." However, Hank Willis Thomas, an artist who argued that it's time for a canon
for co-creation (which he calls a new form of collaboration), defined it as "diverse
and complex authorship."

Not only does co-creation inspire self-reflection, it also suggests shifts in power.
Co-creation is often closely connected to social movements. Jenny Lee, of Detroit's
Allied Media Projects, said that it could be a way to describe "the need for naming
multiple visions and multiple people, in an organizing sense," to move away from
the idea of the "single charismatic leader."

Co-creation is often used to describe a collective working together within a community or across communities. The term can mean professional media makers working
with nonprofessionals—with people who in other contexts would have been clas-
sified as either subjects or audiences. Co-creation, our interviewees agreed, requires
strategy and frameworks.

Co-creation is widespread in the digital and computational arts. In a 2021 survey
of artists using technology as a creative medium, the National Endowment of Arts
highlights the need to recognize the fluidity and hybridity of the genre.[4]

CareForce (2017) is a multimedia public art project about caregivers working in the United States. Used
with permission from Marisa Morán Jahn.

Quite a few of our interviewees said they have noticed the term *co-creation* becoming more prominent in the vocabulary as people search for a way to describe these emerging (yet age-old) sets of practices. "Co-creation isn't a term that I use, but I understand the principle of it," said Anishinaabe filmmaker and artist Lisa Jackson, "In the Indigenous film community, there's a lot of talk about . . . 'reciprocity.' That is a word that's used a lot." Assia Boundaoui, Algerian American journalist and filmmaker, noted that she'd been hearing the term *co-creation* more in recent years and said, "I finally found a word for something that we had been thinking about for a long time, and I'm very interested in telling stories from inside of my community."

But using those frameworks, it's possible for people to transcend the bounds of individual achievement and discover a greater whole, said Opeyemi Olukemi, former vice president of digital innovation at American Documentary's *P.O.V.* at the Public Broadcasting Service (PBS): "Co-creation is the ability for people to humble themselves and let their guard down, to create something that they cannot do by themselves. . . . Co-creation to me is the attempt to reach this whole mind state, where people also understand that everyone has a piece of the collective puzzle. And jointly, and only jointly, can we create something that is truly revolutionary and meaningful."

Jennifer MacArthur, documentarist and media strategist, said, "The end goal is all of us claiming back our humanity, and for where you are, that means different things. Co-creation allows, perhaps, I hope, a way into a process that gets you closer to reclaiming your humanity."

STAGES AND THE SPECTRUM OF CO-CREATION

Co-creation may not be used for all stages of the media-making journey, especially in long-term, ongoing projects. Hank Willis Thomas, a Brooklyn-based conceptual artist, said that one of the biggest challenges of co-creation is understanding that partners and relationships come and go and that in the long view, people may also return. Heather Croall of the Adelaide Fringe described co-creation broadly as moving from the abstract to the specific, from divergent to cohesive systems, requiring different skills and relationships along the way. So too, Ethan Zuckerman, a professor of communication and public policy at the University of Massachusetts at Amherst who has started a new research center called the Institute for Digital Public Infrastructure, described his process as iterative and changing players as the needs require. The Spectrum of Collaboration graphic suggests how co-creation sits in the range of a joint discovery process in which no partner is driving but rather all are copilots.

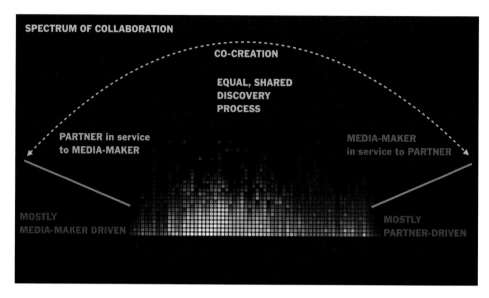

Spectrum of Collaboration Graphic by Katerina Cizek. Courtesy of Co-Creation Studio.

AUTHORSHIP: THE ONE AND THE MANY

Co-creation helps to unpack the myth of the single author. That concept is, historically, a fairly new construction linked to ownership and private property. It is also linked to ideas about the individual. Co-creation suggests instead a distributed notion of authorship (and authority) and creates open spaces for many voices.

To be sure, the category of the author has multiple functions: attribution, authority, credibility, and responsibility. Certainly, all remain relevant for scientific as well as political discourse. This is neither objectively better nor worse than co-creation. Authorship is more visible, more saleable, and more heavily promoted and has affordances that are particularly relevant in the digital age, an age wherein technology can make counterfeiting images and information simple and tracing sources difficult.

However, the inherited historical notion of the author and the positioning of it as the cultural default has come at the expense of co-creation, especially at a moment of digitally enabled cultural change and socially mandated reassessments of business as usual. We say this fully aware of the continuing importance of authorship and attribution at a time of a breakdown in social trust and consensus regarding reality.

The idea of the author is strongly tied to concepts of the self, the ego, and individual will. Films and media projects, for example, are by default considered authored by one person. The funding, production, and distribution models are set up to credit and acknowledge one director or at best two codirectors with author-ship of the work. The entire media industry continues well into this century to be built around these principles, which film, media, and journalism schools teach and replicate.

Co-creation, by contrast, enables alternatives to the way stories are extracted in this industrial model. In a 2019 media production guide to working with Indigenous communities and cultures called *On-Screen Protocols and Pathways*, Marcia Nickerson states, "'Ownership' over stories or Indigenous cultural property is a concept that extends beyond the individual to the community. In obtaining proper consent keep in mind too that there are nation stories and rights, community stories and rights, and individual family stories and rights. When you are contemplating the use of oral histories, understand that there may need to be some limitations of where copyright applies and you may want to consider 'shared authorship' or 'co-creation' credits with community members."[5]

The languages and beliefs that bind human beings and the narratives that serve as our cultural operating systems all have taken form through the long-term active participation of people from across the social spectrum. The dynamic and radically inclusive nature of this process is all too often occluded by the dictionaries, sacred texts, and compendiums of stories that have been extracted by cultural arbiters, sys-tematized, and handed back as dominant culture.

The Rigveda of ancient Hinduism, the holy books of Shinto in Japan, the Tao Te Ching, the Upanishads, the Buddhist Sutras, and the Jewish and Christian scriptures all have undergone a similar process of slowly accreting from the activities of diverse authors and peoples and being re-presented as authoritative texts. The same principle holds for epic literature in the form of the Manas of Kyrgyzstan, *The Mahabharata*, the Tibetan Epic of King Gesar, and the Homeric epics of classical Greece. We tend to celebrate their crystallization as texts but fail to acknowledge their emergence from a dispersed and participatory cultural space. It is easy to miss the nuanced and delicate social interactions that gave rise to culture in the first place, distracted by the institu-tional edifices, hierarchies, and myths of authorship that have been superimposed upon them.

Over the long haul, the practices that this book describes as co-creative have at times been understood as basic modes of sociality and group survival while taking their collaborative dynamics for granted. Terms such as *barn raising* and *quilting bee* say more about what was worked on rather than how. The *Oxford English Dictionary* describes the range of activities for which the term *bee* has been used, alluding to

Mural depicting King Gesar of Ling. Photo by Andreas Gruschke. Image under Creative Commons.

the basic social contract at its core: "A meeting of neighbors to unite their labours for the benefit of one of their number; e.g., as is done still in some parts, when the farmers unite to get in each other's harvests in succession; usually preceded by a word defining the purpose of the meeting, as apple-bee, husking-bee, quilting-bee, raising-bee, etc. Hence, with extended sense: A gathering or meeting for some object; esp. spelling-bee, a party assembled to compete in the spelling of words."[6] The *Oxford English Dictionary* also notes the mid-nineteenth-century usage of the term *lynching bee*, a powerful reminder that not all forms of collective action are benign.

The people of Gee's Bend, Alabama, an isolated community enclosed on three sides by a river, offer a striking example of the power of co-creation through bees. The small predominantly African American and Native American community was impoverished as a result of a history of land theft, plantation slavery, and sharecropping but began to be restored to local control with assistance from the US Farm Security Administration in the 1930s. The ensuing transformation was documented by Farm Security

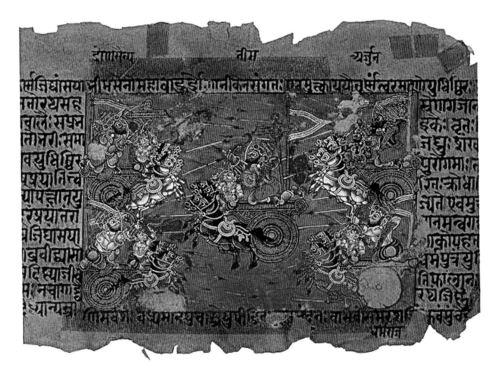

The Mahabharata *is an epic Sanskrit poem, passed down first as a collection of popular stories and performances before being written down in a collected text. Image under public domain.*

Administration photographers Mary Post Wolcott, Dorothea Lange, and Arthur Rothstein, putting Gee's Bend on the map as a successful New Deal experiment. But it was the community's women who drew the world's attention, thanks to their quilt-making. Drawing on inherited African American and Native American textile traditions, their distinctive quilts were exhibited at the Whitney and Philadelphia Museum of Art and many other museums and galleries along with work from the Freedom Quilting Bee in neighboring Alberta, Alabama. While the art world and collectors clamored for the names of individual artists, the women of Gee's Bend held firm to their communal identity, eventually made official in 2003 under the name the Gee's Bend Collective.

The Gee's Bend quilts, like the countless barns raised by neighbors across North America, attest to centuries of co-creation as communities pooled expertise, labor, and commitment in ways that cemented social bonds and passed knowledge from one generation to the next. Co-creation takes place in social spaces, regardless of where in the world they happen to be located (despite this book's generally North American

Women from Gee's Bend work on a quilt, 2005. Photo by Andre Natta. Image under Creative Commons.

orientation), in the co-creative story traditions of Iranian coffee cafés and in the Marrakesh night market as much as in the Iroquois tradition of making cornhusk dolls to help maintain community planting and harvesting skills, as Amelia Winger-Bearskin reminds us (see "In Conversation: Decentralized Storytelling," chapter 5).

Unfortunately, just as with cultural belief and epic narratives, the significant work of the many remains largely invisible, washed from the historical record; nowhere is this truer than in the case of marginalized people. Meanwhile, the attempts of a few individuals to formalize this dispersed labor—to reaggregate, package, and label it—tend to live on.

The turning point that eclipsed co-creative practices was coincident with the industrialization of media that dominated the late nineteenth and early twentieth centuries. This development amplified the Romantic era's notion of the author as creative genius and transformed it into a business model, an instrument of power, and—from Thomas Edison in that era to Apple's Steve Jobs a century later—a brand.

A barn raising on the farm of Daniel and Matilda Dickerhoff in Stafford Township, Indiana, circa 1900. Photo by J. E. Curtis. Image under Creative Commons.

In the visual arts, celebrated masters such as Rembrandt, Rubens, and Rodin often created their work with teams of unnamed assistants, as Jeff Koons does today. But the nineteenth- and twentieth-century biographies of these artists, like the investment markets that fed on their work, opted for a highly selective notion of attribution.

The late nineteenth century also marks the birth of mass media through the industrialization of the newspaper, film, and later broadcasting industries. Michael Schudson's history of the American press chronicles the period's transformation as family-run and community-centric newspapers became overshadowed by national syndicates answering to stockholders rather than communities.[7]

In the case of film, after a few initial years in which amateurs and craftsmen experimented with the medium and even the *Sears Roebuck & Co. Catalogue* encouraged its customers to buy film equipment and organize community screenings, Thomas Edison intervened with lawsuits, essentially claiming the medium as the property of the Edison Manufacturing Company. Although Edison did not prevail in the long run, the company's actions opened the door for the expansive, hierarchical, profit-centric entertainment industry that was in place in the United States by the time of World War I.

Industrialization was nurtured by technological advances (often stimulated by the period's wars) and went hand in hand with population migration and urbanization. But while industrialization has been interpreted conventionally as progress, it also resulted in collateral damage, overshadowing its artisanal and sometimes co-creative counterparts. Nonspecialized community-centric endeavors—be they architectural (barn raisings), artistic (quilting), or media (community newspapers)—were recast as folkish, amateurish, and premodern. Particularly in the Western world, those views shaped the further evolution of concepts of cultural ownership and intellectual property through, for instance, patent law and copyright.

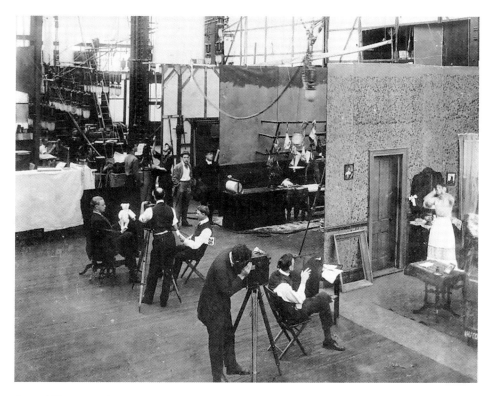

Several films in production, circa 1914–1915, at Thomas Edison's Bronx studio. Image under public domain.

Collective authorship, or even the collapsed authorship of co-creative models, breaks down these systems of individualism, ownership, and property. "If one makes something uncommodifiable, how does that recast things, and allow different types of relations to come out?" asked Chi-hui Yang of JustFilms at the Ford Foundation. These tensions do not mean that co-creation is necessarily antithetical to existence in a marketplace, but it might have to negotiate its role there in distinctive ways.

Almost simultaneously with the Industrial Revolution–era turn toward concepts such as individual authorship and production, in fact, a rich lineage of pointedly collective and collaborative counterefforts was born. Co-creation has continued under various names and programs, from Dziga Vertov's Soviet film collective through Canada's Challenge for Change project at the National Film Board and Bolivia's Grupo Ukamau to countless artists' collectives, radical 1960s film collectives all around the world, transdisciplinary initiatives and more as well as at the community level. Some of these

The Soviet filmmaker Dziga Vertov (1913), founder of the Kino-Eye filmmakers. Image under Creative Commons.

practices have strong roots in theater, such as Augusto Boal's Theater of the Oppressed in 1960s and 1970s Brazil, which positioned theater within people's movements by breaking the fourth wall between actors and spectators. "We are all *actors*: being a *citizen* is not living in society, it is changing it," as Boal wrote in a 2009 World Theatre Day message.[8] Devised theater is another system dedicated to collective creation, and verbatim or documentary theater draws on scripts built from interviewing real people. Indeed, a surprising number of interviewees in our study had backgrounds in theater, dance, comedy, or performance, which might have primed them for an interest in co-creation in other media.

Still from Grupo Ukamau's Blood of the Condor *(1969). Image under Creative Commons.*

"There's a particular affinity, for some reason, between our documentary filmmakers and the theater-program fellows," said Tabitha Jackson, director of the Sundance Film Festival. "When we're here on the mountain together, there's also always some crazy stuff that comes up. There are so many examples of using these different disciplines to come up with a new or an expanded vocabulary of cinema."

The scholar Henry Jenkins, drawing upon his long interest in fan communities, famously identified the phenomenon of "participatory culture" in 1992, referring to practices in which fans act as bottom-up contributors and creators rather than just passive consumers of unidirectionally aimed authored works.[9] These fandoms challenge the authority of the author; they mix and remix characters and situations so that they take on a life of their own. The internet has facilitated that kind of participatory fan culture as well as countless other co-creative (and sometimes co-destructive) practices through the web and social media.

Indeed, the web, especially the nontextual social web such as YouTube and TikTok (video) and Clubhouse (audio), has reconfigured individual and collective senses of time and space, introducing an immediacy and a complex feedback system to this cacophony of collective, multiple, and collapsed authorships. Digital culture extends connections outward while seemingly rewiring people's own inner circuitry. However, as digital empires have consolidated in the last decade, corporations have found ways to recommodify and reintegrate this participatory content into a supercircuit for their own increased profit. Digital culture has opened up new opportunities for connecting, contributing, collaborating, and co-creating. These are complex and still evolving terrains where, compared to some media forms of the past, ordinary people can rightly claim a more active part in their creation. But the nature of that role—the limits of agency and the constraints of structure—elicits different views, from utopian to dystopian.

"What kind of problems are we solving? Who owns these problems? What kind of needs are we trying to meet? Where is the context for these questions to be posed?" asked Carlos Martinez de la Serna, program director at the Committee to Protect Journalists.

Correcting for the domination of the single-author model in the past cannot mean merely crowdsourcing creativity and media without regard to whose voices gain traction. Co-creation should open spaces between makers, disciplines, and the communities formerly known as subjects and the people formerly known as audiences. Authorship and co-creation have different strengths, and there is a need for both.

A still from Tatsuniya *(2018), a collaboration between artist Rahima Gambo and ten girls from the Shehu Sanda Kyarimi government school in Maiduguri, Borno State, Nigeria. Used with permission from Rahima Gambo.*

WHY CO-CREATE? WHY NOW?

The big question is why co-create? Why embrace a mode that is often underrecognized, perceived as messy and slow, and institutionally unsupported? Why now, when media making is already difficult to fund, sustain, and share? In our research, we kept running across five central reasons why people felt co-creation was worth the effort.

1. *Co-creation helps us navigate the uncharted changes sweeping the planet*, including technological, environmental, cultural, political, and economic upheavals. All of these are intertwined in patterns that legacy twentieth-century models are unequipped to handle, while co-creation can draw on wider perspectives on intersecting issues.

2. *Co-creation confronts power systems* that perpetuate inequality and offers alternative, open, equitable, and just models of decision-making rooted in social movements.

3. *Co-creation can lean into complex problems* with a commitment to finding solutions at and with peoples living at the local level.

4. *Co-creation deals with time differently* by insisting on responsiveness while expanding the time frames in which we consider and measure consequences.

5. Finally, *co-creation is part of an ecosystem of practices across many disciplines that redefine the public good, civic trust, and the commons.* How do we share the world with each other, human, nonhuman, and perhaps beyond human?

"People are beginning to realize that this superindividualistic view and process of living are quite damaging to the individual and to the planet," said Opeyemi Olukemi, formerly of *P.O.V.*

"My experience is more anchored in the social movements," said Monique Simard, a veteran producer and cultural development executive in Quebec. "And because I've been around for a while, I realize how social progress has been achieved by not only co-collaboration, co-working, solidarity, but also . . . by points in history that merge together and make you do the 'big leap.' And after that you can't go back in history or, at least, you're better positioned not to go back in history. All this energy and dynamics come together and make change happen."

Let's take a closer look at each of these motivations for co-creation.

During a think tank session, Sudanese and Eritrean asylum seekers analyze problems from residents of the neglected neighborhood of South Tel Aviv. Photo by ArtPort Gallery for Ghana Think Tank.

Ghana ThinkTank Mobile Unit, Corona Queens, New York (2011). Maria del Carmen Montoya and John Ewing push the Ghana ThinkTank mobile unit through the streets of Corona, Queens. Photo by Christopher Robbins. Used with permission from Ghana ThinkTank.

I. NAVIGATING UNCHARTED TERRITORIES

We look at the present through a rear-view mirror. We march backwards into the future.
—Marshall McLuhan, *The Medium Is the Message* (1967)

Co-creation helps us sketch routes and maps for use in the uncharted territories of the twenty-first century, draw on our collective intelligence, and shed old legacy models that have become irrelevant.

"What's happened post–World War II with consumer capitalism, hyper-consumerism, it's chipped away at our will, public will, and interest for collaboration," said Michael Premo, a documentarian and artist based in Brooklyn. "Now, the explosion of technology is reminding us of the collaborative nature of social organization that we haven't maybe thought about for 100 years, that can really kind of push us into a new place."

New technologies, then, both afford and demand new ways of working. Yasmin Elayat of Scatter, a hybrid start-up, next-gen creative company told us that "co-creation for me is pretty central to anything that I call interactive. . . . Any work that I would do, defined as interactive, I believe, needs to be actually co-created with the audience."

We also need new frameworks for understanding what is already happening. Patricia R. Zimmermann and Helen De Michiel have offered the term "open-space new-media documentary" in their book of the same title. They state that "earlier theoretical models focusing on representation and aesthetics cannot account for new collaborative documentaries sited in the lived experiences of particular individuals in specific places."

Some interviewees have observed that "ordinary people" have bypassed the institutions and the systems that are now obsolete, finding their own possibilities of technology and networks.

Co-creation provides the grammar and the vocabulary for these uncharted territories, or what cognitive scientist Kristian Moltke Martiny of the University of Copenhagen called the "third space," and many others interviewed defined as in-between space.[10] "It's a transcendent idea going beyond what the discipline entails," said Martiny, one that adds "its own structures, its own narratives, its own self-awareness, its own methods or processes to a third space together with one of the other disciplines that you're co-creating with. Which also means it's an unknown. It's an explorative process, typically."

Zero Days VR (2017), produced by Scatter, directed by Yasmin Elayat. Used with permission from Yasmin Elayat.

2. CHALLENGING INEQUALITIES

The most common way people give up their power is by thinking they don't have any.
—Alice Walker

Co-creative processes challenge and bring transparency to the power dynamics of relationships entrenched in the legacy models of professional production—between the people formerly known as the makers, the subjects, and the audiences. A new space emerges for equitable, inclusive, and democratic practices. For people with access to institutions, on a personal level co-creation means leveraging that power. As Salome Asega, artist and director of NEW INC. at the New Museum (New York), commented, "I have maybe been granted access to certain resources, so I partner a lot with community groups or subcultural movements. . . . I'm always thinking about, 'How do I bring people into the spaces?' Once I have an invitation, how do I extend that invitation?"

This process of redistribution and sharing access, creation, and decision-making happens at a microlevel within productions, but it extends out to a macrolevel of movements, organizations, and broader social relations.

"Beyond the work itself," said Michael Premo, "co-creation is a model of reorganizing society in a way that really favors collaboration and reorganizes our relationship to hierarchies in a way that kind of models ways of how we relate to each other."

"Co-creation is generous," said Monique Simard. "It's using the best of everybody's skills. It is, I would say, democratic. It shatters a notion of one author, one person owning 100 percent of a product. . . . It goes against exclusivity. It goes against all these notions that were very, very valued by neoliberalism."

In the twenty-first century, neoliberalism's values have been under broader attack by large intertwined social movements that flatten hierarchies, broadly called "horizontalism." As scholar Marina Sitrin defined it in an article in the magazine *NACLA Report on the Americas* in September 2014, "*Horizontalidad*, often translated as horizontality or horizontalism, was first used by the movements that emerged in Argentina in the wake of the 2001 economic crisis. It has since been adopted by social movements around the globe, and is used to describe new forms of social relationships that are developing in place of traditional methods of political organizing . . . from Spain, Greece, Bosnia, and Brazil, to the U.S. Occupy movement."[11]

Decentralized structures have become a defining feature of movement building in this century, as was especially visible in 2020 from Black Lives Matter in the United States and elsewhere (which brought historic numbers of people into the streets) to the prodemocracy protests in Hong Kong, antiausterity activists in Chile and Lebanon, and anticorruption demonstrations in Haiti and Iraq. Often preferring the term *leader-full* to *leader-less*, the members of these diverse movements agree that without formal leadership, they are less vulnerable to being attacked from the top down (as with the assassinations that afflicted the 1960s US civil rights movement) and benefit from the flexibility of decentralization. In Hong Kong, activists embraced Bruce Lee's motto "Be Water" to express this Tao-inspired strategy.[12]

Lam Thuy Vo named the moment in a 2020 article titled "Adversarial Experiments Everywhere": "The Internet is filled with chatter around divestment: divesting from police to social services, divesting from Bezos to Black-owned businesses, divesting from incumbent politicians to everyday people who are running for office, some for the first time. Maybe it's a good time for us to push out new projects that model a different way of being."[13]

A protest in Argentina against the banks in 2002. The large sign reads "Thieving banks—give back our dollars." Photo by Pepe Robles Cacerolazo. Image under CC BY-SA 3.0 license.

At the core of co-creation, equity and diversity are fundamental to addressing power dynamics within projects and at deeper structural levels, as laid out in the searing call to action by media consultant Kamal Sinclair in her research project *Making a New Reality* (2018): "Emerging media cannot risk limited inclusion and suffer the same pitfalls of traditional media. The stakes are too high. Together, we must engineer robust inclusion into the process of imagining our future."[14]

Power dynamics are always present, but co-creation can help name them, which can be the first step toward finding new systems of relations.

3.   CLIMATE CRISIS AND COMPLEX PROBLEMS

Without stories of progress, the world has become a terrifying place. The ruin glares at us with the horror of its abandonment. It's not easy to know how to make a life, much less avert planetary destruction. Luckily, there is still company, human and not human. We can still explore the overgrown verges of our blasted landscapes—the edges of capitalist discipline, scalability, and abandoned resource plantations. We can still catch the scent of the latent commons—and the elusive autumn aroma.

—Anna Tsing, *The Mushroom at the End of the World*

The complex problems that humanity faces in the twenty-first century are too big for old, legacy systems. As we described in the introduction, the COVID-19 pandemic required unprecedented global scientific cooperation to tackle such a massive complexity. Overwhelmingly, co-creative veterans name the climate crisis as a top priority to be tackled from the ground up. Complex problems need large teams, and solutions are often found in the communities that are impacted most by the problems. Julia Kumari Drapkin of ISeeChange connects dots of data with stories, from the ground up to the sky, using NASA's satellite images: "Climate change is so large

Production image from Marina Zurkow's Mesocosm (Wink Texas) *and* Hazmat Suits for Children *(2012). Software-driven animation and Tychem TK sculptures. Installation documentation, bitforms gallery. Photo by John Berans. Used with permission from Marina Zurkow.*

and big and coming at us from such large amounts of time and space," she told us. "We need to be drilling down into the specifics of how a community is experiencing it and what's causing it. It's that tangible community context that allows solutions to happen, that allows the journalism to happen."

"Will it really matter what we create, whether it's a project or an initiative, if we don't have clean air to breathe?" asked Opeyemi Olukemi. "If we don't have water to drink? If we have a series of superbugs that start to kill off entire populations? Not just to create, but to be responsible and have people realize that we are entering new territory and that this is a possible way to help address and stem the damage of what is coming down the pipeline."

These crises intersect with the trajectory of surveillance capitalism and growing global authoritarianism. They are bound to the gross exploitation of social media platforms and coded inequities of algorithmic systems. The anthropologist Anna Tsing has developed the concept of collaborative survival, the idea that the human capacity to persist as a species is linked, entangled, and rests upon the health of other species. Co-creation can likewise provide a set of methods and techniques to pursue that hope and to distribute resources and governance more widely by considering co-creation itself as a process between species and with nonhuman systems.

Extreme weather conditions are becoming more common as the climate crisis accelerates. Image under Creative Commons.

4.   DEEP TIME

What time is it on the clock of the world?

—Detroit writer and activist, Grace Lee Boggs, as quoted by Robin D. G. Kelley in the foreword to *Living for Change: An Autobiography*

Co-creation stretches conventional notions of time. It may be a term that seems to be currently entering the zeitgeist, but it refers to millennia of human practice. It can also describe processes that may be perceived as long, messy, and not defined by what they can deliver at the outset. Co-creative processes acknowledge that trust, relationships, and democracy take time. The practices challenge the perceived efficiency of capitalistic models, the templates of production that aim to keep costs down for maximal profit. Co-creation defies the bottom line. Co-creation acknowledges deep time. It accounts for artistic projects that aim to unfold over decades and even centuries, projects that may integrate the time frames of rock faces and forests.

This expanded sense of time allows for back-and-forth responses among a wide range of actors. Yet on a parallel track, because co-creative projects have multiple goals and multiple outcomes, they can also be suited to exploring rapid responses to emerging issues. The global COVID-19 lockdowns have highlighted the stretched and compressed accordion of pandemic time at collective and domestic psychological scales. We need processes that allow for both telescopic vision into the far-off past and future and the ability to urgently adapt to sudden developments.

5.   THE COMMONS AND REBUILDING PUBLIC TRUST

To those who are awake, there is one world in common, but to those who are asleep, each is withdrawn to a private world of his own.

—Heraclitus, fragment 95, *Heraclitus on the Two Antithetical Forces in Life* (ca. 535–475 BC)

The traditional term *the commons* refers to public spaces, including cities and resources, but can also include the media, communications platforms, and narratives. Co-creation can help build and restore the commons at a time of heightened polarity and privatization. Co-creation seeks to find shared ground in unlikely places to rebuild dialogue and a mutual vision for governance.

In the *HIGHRISE* project at the National Film Board of Canada, an agency of the federal government of Canada, Katerina Cizek and associates worked for over five years in a privately owned high-rise in suburban Toronto (there are over a thousand such

The Flammarion engraving called Wheel of Time *is by an unknown artist. It first appeared in Camille Flammarion's* L'atmosphère: Météorologie populaire *(1888). Image under public domain.*

midcentury high-rises in the city). The meeting spaces and participatory media work-shops that were created there allowed for people from many divergent backgrounds, political stances, and religions to reimagine their shared place by, for example turning abandoned tennis courts into playgrounds, neglected yards into community gar-dens, and parking lots into pop-up markets. The resulting media outcomes—and there were many, from community bulletins to internationally acclaimed documentaries—became poetic, compelling tools to share with government actors, urban planners, and private owners. Together, these collections insisted that residents' perspectives needed to be included in the Tower Renewal Plan for the city and for the province of Ontario. The initiative helped shape policies at multiple levels of government. But also, through its extended public presence, *HIGHRISE* as a whole helped reshape Toronto's own understanding of itself as a growing vertical, digital city and what that means for the commons.

Similarly, the Civic Signals group, founded in 2018, draws on principles of urban planning to improve digital public spaces (by reforming voting, for example, and managing harassment online) around the world.[15] Civic engagement and the arts are one and the same for artist Hank Thomas Willis, whose 2020 project, *For Freedoms* is decentralized across all fifty US states at the local level, from billboards to town halls to large-scale art projects. And media scholar Henry Jenkins advocates for the civic imagination, meaning "the capacity to imagine alternatives to current social, political or economic institutions or problems." He sees this being done through bringing together diverse players at the ground level around a common issue.[16]

In journalism, Carlos Martinez de la Serna, in *Collaboration and the Creation of a New Journalism Commons* (2018), a Tow Center report, calls for a commons that "can be described as an intricate resource system, functioning under an open access regime, with both local and global dimensions, and hosting all components and social activities pertaining to journalism."[17] The commons are connected to the construction of "truth" in journalism, as danah boyd argued at a 2019 Knight Media Forum, saying

HIGHRISE: One Millionth Tower *(2011), sketching ideas on overlays over photographs. Used with permission from the National Film Board of Canada.*

The resulting collaged animated version of the film in which residents imagine a farmer's market in the parking lot on weekends. HIGHRISE: One Millionth Tower (2011). Used with permission from the National Film Board of Canada.

Henry Jenkins (center, in turquoise shirt), attending a civic imagination workshop in Bowling Green, Kentucky (2017). Photo by Gabriel Peters-Lazaro. Used with permission from the Civic Imagination Project.

"truth is fragmenting, but it's not just about information, it's about community."[18] She insisted that to counter this epistemological fragmentation, decision makers in journalism need to build community and deal with not only content but also context.

Co-creation is about holding ground for the commons and building new platforms of public trust, though not necessarily in traditional spaces. As Cynthia Lopez, executive director of New York Women in Film and Television, suggested in our interview, "Years ago, we would think that public media has to be on public television, right? Now, we understand that there are pieces that, whether they're on Netflix or they're on HBO or they're on Showtime, they still have that public-media agenda, but the distribution outlet is not a public one." Similarly, the American Journalism Project, announced in 2019 with a focus on reinvigorating local news, is supported mostly by private funding.[19]

Many of our interviewees insisted that the larger project of public media cannot be left only to private companies, foundations, and philanthropists, noting the importance of public funding.

Ana Serrano, president of Ontario College of Art & Design University, saw the now-defunct Sidewalk Labs experiment in Toronto as a pivotal example of this dilemma over the commons. Sidewalk Toronto was a "smart city" proposal that would have remade an entire district of the city's waterfront. It was intended as a public-private partnership among three levels of government and Alphabet, the parent company of Google, one of the largest transnational conglomerates of our times.

"We've heard for the last twenty years," Serrano said, "this narrative that suggests that governments don't know how to do anything and that they're bad at doing things and that they're not the ones that can make these decisions. They're 'not innovative.' I think that's been an unfair narrative for many, many years, and I think that's the dominant feeling of many citizens. Now, you have corporations saying that they can explicitly make these better decisions in partnership with government."

Alphabet abruptly abandoned Sidewalk Labs in May of 2020 after much pressure from local community groups for transparency and accountability, although the company cited economic uncertainty as the reason for the pullout.[20] Yet in the very same week, former Google chief executive Eric Schmidt was announced to lead a panel planning New York's post-COVID-19 tech infrastructure. The community-based pushbacks soon began, taking up again the hard work of rebuilding resistance in new jurisdictions.[21]

Now, more than ever, to ensure greater capacity for co-creation built with equity and justice, governments and public organizations need to update their digital infrastructures and cultures to fully recognize and respond to the implications of public

Smart city is a term used to describe an approach that prioritizes tech solutions at the expense, critics argue, of public interest, good governance, transparency, and citizens' privacy. Photo by Jai Jinendra. Image under Creative Commons.

space, civic trust, and the commons in a digital age. This would include examining and addressing

- who owns and runs data,
- who owns the algorithms,
- the ways they are programmed and managed in cities,
- who owns and governs media and tech platforms, and
- how to rebuild public trust in a time of fragmentation and heavy digital commodification.

At the website *Making a New Reality*, Kamal Sinclair describes the challenges of trying to identify how to create commons or public space in emerging media. Whereas traditional film and television were based on a well-defined system of local screenings or a single-channel broadcast, the new communication architecture is more

Making a New Reality. Used with permission from Kamal Sinclair.

dynamic, pervasive, and complex. Sinclair asked numerous people in the field how to create public media in this emergent system, and they pointed to the need to understand the commons in multiple layers beginning with the basic infrastructures of the internet itself through to complex accessibility issues of various online platforms. Sinclair describes those aims in these stages:

- Web 1.0—Public media goal is to get everyone access.
- Web 2.0—Public media goal is to get everyone access to safe and equitable platforms.
- Web 3.0—Public media goal is to increase diverse representation, leadership, and participation in building the new internet.
- Web 4.0—Public media goal is to ensure diverse minds and communities are leading in the design of our AI future.[22]

Co-creation can help provide a pathway to those Web 3.0 and 4.0 goals, being mindful of the hazards to the commons that digital interaction can introduce or heighten, including misinformation, trolling, appropriation, data insecurity, and other forms of abuse. How do creators and the public imagine a future that is, by contrast, inclusive, equitable, transparent, and sustainable?

CO-CREATION IN THE WORLD

Co-creation and related practices are proliferating across many disciplines and sectors. We intend our definition to allow for conversations across many divergent fields, areas of knowledge, and traditions. The co-creation wheel of associated practices diagram illustrates the connections.

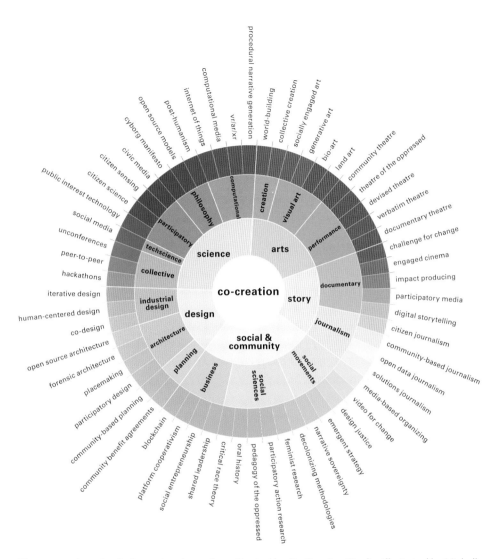

The co-creation wheel of associated practices. Created by Co-Creation Studio. Illustrated by Michelle Hopgood.

In its effort to center on communities and question conventional expert knowledge, media co-creation shares commonalities with practices in many other fields. Some embrace human-centered practices primarily for profit (e.g., in product development) while other more justice-based practices can challenge power dynamics. In many professionalized fields, community-based co-creation processes are called *participatory*. Architects, for example, may work with the members of a cooperative for the hearing impaired in order to understand and co-design an appropriate building for them to live in. Matthew Claudel, an architect at MIT, calls this "open-source architecture" and defines it in opposition to the "starchitecture" that valorizes a singular rockstar-like architect-author.

"Open-source architecture is the way that we've built architecture for millennia," said Claudel. "It's this vernacular idea of architecture, that people construct context for daily life. It has beautiful results like the adobe dwellings in the southwest of the United States that are very different than the wood-slat houses that would have been built in the northeast. And everything in between and beyond. Vernacular architecture, it's really an expression of community, culture, climate and place."

These approaches also extend to urban and rural planning in a critical practice often referred to as *place making*, which emphasizes processes that bring communities and residents into planning.

In social sciences, health, and other research practices that involve fieldwork, participatory methods have been codified as participatory action research (PAR). Informed by feminism, PAR suggests that even designing the research question in the first place requires deep listening at the community level. PAR requires researchers to ask what the community needs to be addressed, as opposed to what an outside expert wants to research. Elizabeth Miller, a documentarian based in Montreal, said she believed that PAR "means curating a space instead of presenting a research question. It often means sharing skill sets and beginning there. It often means extensive dialogue before you even begin the process of research."

Participatory methods and even the term *co-creation* have also cropped up in the fields of organizational and business management. Management academics such as C. K. Prahalad and Venkat Ramaswamy have used the term to describe "the joint creation of value by the company and the customer . . . allowing the customer to co-construct the service experience to suit their context."[23] In this model, the customer is always right—or at least, the customer is centered and asked to contribute to and evaluate the creation of a product or service. However, in more progressive interpretations it can also involve questioning the single bottom line of profit and creating the framework of a double or triple bottom line that also measures social and environmental impacts.

In other disciplines, such as engineering, industrial design, and software design, evolving sets of similar practices are called human-centered design, or co-design. These concepts are based on the insight that engineers cannot design useful products in the vacuum of their labs without input from their potential users. Although popular, these practices do not guarantee democratic design or fair outcomes "if you're just mining users for ideas and then you create the product and sell it back to them," said Sasha Costanza-Chock, author of *Design Justice*. "Yes, you've made a product that's more usable, but it's an exploitative relationship with people." In response, a relatively new practice, design justice, which gives Costanza-Chock's book its title, prioritizes the voices and lives of people and communities who most risk adverse effects from design decisions.[24] Importantly, the questions of who chooses the design problem, how can benefits be most widely shared, and how can exploitation be minimized remain.

Workshop participants prototype in Making the Breast Pump Not Suck *(2018). Used with permission from Catherine D'Ignazio.*

Babitha George, a partner in Quicksand, a design firm in Bangalore, said that in her community-based water sanitization work in India, she finds parallels between local social movements and many of the Western-defined design practices she draws on. She cited her friend and mentor, Vishwanath, whose water conservation work, according to George, has proved that "solutions or answers have to be found within communities. And therefore this notion that everything can be scaled is not true at all." Vishwanath left a government engineering job to look for nonengineering solutions to water conservation and now works in deep partnership with local collectives. As Zenrainman, he has also become a popular YouTube personality, with videos that share and demonstrate his methods and solutions.

At their best, these community-based methodologies show that designs and solutions can be found in traditional and local knowledge and in grassroots movements, especially when thoughtfully combined with the use of accountable research methods, open technologies, and appropriate engineering. In our interviews, however, opinions ranged about whether these co-creation models truly flatten hierarchies or risk deepening existing inequities. Some practitioners reason that community-centered projects can at least make power dynamics more visible and explicit. Others call for researchers to relinquish power to communities in order to have authentic meaning and a positive impact.

IN CONVERSATION: CO-CREATION AND EQUITY—FIVE MEDIA MAKERS OF COLOR SPEAK OUT

An excerpt from a conversation between Juanita Anderson, Maria Agui Carter, Thomas Allen Harris, Maori Karmael Holmes, and Michèle Stephenson.

Inherent in the co-creation process is a political yearning and narrative turn toward a more ethical form of storytelling. This is undoubtedly an auspicious moment. However, for media makers of color, there's one vital concern. While it has been acknowledged that the co-creation of stories can be traced back to early civilizations, there has not been an explicit recognition of the specific ways media makers of color have shaped and inspired contemporary co-creation practices emerging from accountability to their communities, the drive for more complex representation, and the necessity of creating media with less financial resources than their White counterparts. Oftentimes, media makers of color, especially those documenting urgent political issues, are faced with increased demands of proving their professional media making expertise within the field, acting as cultural interpreters between their communities

Clockwise from top left: Maori Karmael Holmes, Juanita Anderson, Maria Aqui Carter, Thomas Allen Harris, and Michèle Stephenson. Used with permission.

and the field, and ensuring respectful engagement of their cultural communities. This complex navigation has given rise to deeply organic and mindful co-creation processes that predate the increasingly popular use of the term in the media field.

Presenting co-creation within media as a reemergent practice without explicitly acknowledging the long-standing co-creative approaches practiced by communities of color doesn't simply erase their work; it undermines the very tenets of co-creation.

At its core, the idea of co-creation seeks to reconcile systemic power and singular authority. This fundamental principle extends beyond the creative process within media making, compelling all us to interrogate fundamental ideas of ownership, meaning making, and attribution and—if we optimize the potential of co-creation—to realize a more just society. With this larger imperative in mind, this chapter seeks to provide an overview of the long-standing co-creative practices of artists of color. The idea for this work is derived from a breakout group conversation that took place at the two-day symposium Collective Wisdom: Co-Creating Media with Communities, across Disciplines and with Algorithms, organized by the Co-Creation Studio at the MIT Open Documentary Lab. Its content is not meant to be comprehensive, nor does this work represent the multifarious approaches and perspectives of media makers. The function of this chapter is to begin mapping the history of co-creation across communities of color within the United States; to explore noninstitutional power, innovation, and cooperation among media makers of color; and to unpack the uncalculated costs and labor of deep co-creation processes.

The content presented here is based on a conversation structured around a series of prompts that have been edited for length.

INTERROGATING THE MAINSTREAM FRAMING OF MEDIA MAKING

JUANITA ANDERSON: For me, the moment that I questioned what we were doing at MIT was during a panel session where someone commented that they spoke to the community members rather than journalists. And I was thinking obviously, as people of color, we always speak to our communities first. We've been working from a co-creation model for decades—taking us right back to the history of African American, Latinx, and Asian American programming in public television back in the late 1960s. Television series like *Black Horizons* in Pittsburgh, *Say Brother* in Boston, and *Detroit Black Journal*, all launched in 1968, had community advisory boards and community input all along the way in terms of both the content and management of those programs. The failure to recognize this and many other histories demonstrates an issue in current co-creation conversations that hasn't been resolved. There doesn't seem to be a priority to support media makers of color working within their own communities. Are we talking about White media makers going into communities where they have never been before?

THOMAS ALLEN HARRIS: I definitely concur with what Juanita says. It's also important to note that within the larger context of colonialism, there is a certain class of people going in and telling the stories of a wide range of historically marginalized communities. Today because of the economics of long-form documentary production as well as pressure by distributors for diversity of representation, there is increased talk of co-creation, which I think is a response to these changes. And I'm wondering if collaboration is really happening or whether it's a certain kind of blackface or Asian face—a cover to continue to exploit these communities for content and access to funding and distribution in an age of multiculturalism.

A focus of many of my mentors—William Greaves, St. Clair Bourne, Pearl Bowser, Marlon Riggs, Camille Billips—was critiquing dominant modes of representation that served to disenfranchise their communities while simultaneously empowering these communities by providing access to tools, a voice as well as modalities of storytelling, that communities could use for self-representation. This process—and the community building it engendered—was considered in some cases just as important as the final film.

This process of community engagement as part of the production, together with a merging of art and activism, placed many of these projects within the avant-garde. I'm thinking particularly of Greaves's *Symbiopsychotaxiplasm: Take One* or Camille Billops's *Suzanne Susanne* or the early pioneers of Black independent film that Pearl Bowser's work has documented over decades or Marlon Rigg's *Tongue Untied*. Yet, much of this work was not written about or critically appraised sufficiently within the field until years after they were made, if ever. This benign neglect, often due to the work's resistance to certain stereotypical narratives, resulted in its marginalization so that today some can speak about co-creation as something new without feeling the responsibility to find and cite precedence within media makers of color that have long been ignored by the mainstream. The result is a kind of a painful double negation. So, as we revisit or repackage the concept of co-creation, it's important for us to interrogate our power relationships and our motivations vis-à-vis process—community as well as outcomes.

MICHÈLE STEPHENSON: Yeah. I want to affirm what's already been said. I also want to question the current emphasis on co-creation in the field. Is it simply a way to legitimize a space for dominant culture to continue to be involved in telling our stories, a way of keeping some sort of White control and presence over narratives by and about communities of color? I won't get into the whole capitalism versus collaboration model that has existed in media making for a long, long time, but as Thomas mentioned, power is shifting more significantly now in dominant spaces. We can't ignore that this co-creation conversation is gaining all this visibility and weight at a time when people of color are commanding more authorship power. Suddenly, single authorship is being questioned, and talk of co-creation is taking place within the mainstream.

MARIA AGUI CARTER: I couldn't agree more with Michèle, Juanita, and Thomas—such great points. I originally studied anthropology, a discipline predicated on the belief that it is possible for an outsider to parachute into a community and gain access to its culture, knowledge, and stories, then speak for them to the rest of the world. I remember reading Claude Levi Strauss's *Tristes Tropiques*, a book by the much-celebrated French anthropologist, one of the fathers of anthropology, whose memoir I found racist and problematic. As one of the "natives" that he would have talked about, I found him to be so bound by his own cultural biases that he could not truly see, and I questioned his capacity to spend limited time with a community and yet understand it well enough to properly frame the story or ask the right questions. He is someone who spent little time in the field and yet made sweeping pronouncements about how the "natives" thought and explained their societies, their philosophies. I think of the arrogance of his observations and his many books as an "expert" on other cultures, given he was known to have said "fieldwork is a kind of women's work (which is probably why women are so successful at it). I had neither the interest nor the patience for it." It reminds me of documentary filmmaking, when people (usually extremely sympathetic and well meaning) from outside our communities of color choose to speak for us without having put in the equivalent of the research and fieldwork that earns them the right to make those observations. The research we do as filmmakers of color about our own communities often lasts a lifetime and is not just that of lived or shared experience. So many of us specialize not just in the craft of filmmaking but also in reading and researching the arts, the culture, the politics of our own people and our communities as well as those of the larger communities around us.

With regards to co-creation and collaboration, as filmmakers of color making films about our communities, we have always felt that we needed to speak and to listen to those from within. Yet we are not credited by the power structures as experts—instead we are often accused of bias—nor are we often funded adequately to tell our own stories. For instance, in the foundation world in the United States, the last studies I read showed that Latinx organizations across all fields receive about 1 percent of all foundation dollars—that includes everything from direct services to the arts and film. Yet we make up almost 18 percent of the US population, and we now have a very large community of trained Latinx media makers and storytellers. And so it's not just that our stories need to be told, it's important that the world respect, fund, and support our telling our own stories.

MAPPING MEDIA-MAKING PROJECTS: A FEW GOOD EXAMPLES

JUANITA: I want to build on Maria's comment about anthropology, because you think about the work of European visual anthropologists like Jean Rouch, whose ethnographic filmmaking in African societies during the 1940s and '50s was often in direct service to colonial interests. A prime example was *Les Maitres Fous*, in which his narrative framing of Hauka spiritual practices in Ghana were reported to have so incensed African peoples that in some respects he is credited with giving rise to the African cinema movement, which rejected and countered the European gaze on African societies. From the work of Senegalese filmmaker Ousmane Sembène, whose pioneering 1963 film *Borom Saret* is considered to be the first film by an African film-maker to be made on African soil, African filmmakers, at least in the West African countries that I have visited, have always engaged communities in the making of their films—and not just as actors but as production crew members, community advisers, production designers, and the like. So, there are international models that demonstrate people of color calling upon our communities and our families to make media.

THOMAS: Yes, and Juanita, I'd like to build on your excellent point [earlier] about public television. You also had innovative shows that brought community into the studio such as Ellis Haizlip's *SOUL!*, which gave voice to the burgeoning Black Arts Movement that was sweeping across the country in the late '60s and early '70s. I worked closely with Ellis when he was at the Schomburg Center, and he made it his mission to mentor the generations of cultural producers that came after him. Like Ellis, I first began my media career and my training at WNET in the late '80s. My tenure there coincided with the movement around responding to the HIV/AIDS epidemic, particularly as it intersected with the communities of color, queers, and artists. Highlighting these communities became a focus, and because I was in public media, I was able to incorporate these voices within my shows years before commercial television would do so.

I think it's also important to acknowledge the roles that the newsreels of the 1960s played in shaping this landscape as well. Third World Newsreel [TWN] is where I actually took my first class. Cara Mertes [former director of JustFilms at the Ford Foundation] was also in that class. In addition to educating generations of filmmakers, TWN's mission is to create and empower community through media as well as building and maintaining archives of and by communities of color and other disenfranchised communities. Christine Choi [codirector of *Who Killed Vincent Chin*] was one of the instructors. Pearl Bowser [author, director, producer] was another. Both Bowser and Michelle Materre [producer], were really instrumental in making media by filmmakers of color accessible to audiences and building audience support for projects before the idea of crowd funding became a thing. They and TWN were especially helpful in facilitating connections and collaborations between diasporic filmmakers and diasporic communities.

MAORI KARMAEL HOLMES: One of the examples that I can think of is the Scribe Video Center in Philadelphia, which has two projects that I think are good examples. One of the projects is the Precious Places Community History Project. This project is a place-based oral history program that teams neighborhood groups with experienced filmmakers and humanities consultants to make locally authored documentaries that explore the political and cultural history of public spaces in their neighborhoods. Another [project] is called Community Visions, which is a free ten-month video production program for members of community groups wherein participants learn to produce short documentaries about issues of importance to their constituencies. What these projects have in common is that a community organization or an agent within a neighborhood applies to this program and then gets paired with one or two filmmakers to make work about themselves. I'm still sort of learning about the term *co-creation*, but these examples are the opposite of the traditional filmmaker going out in search of community; instead the community is inviting in the media maker, and both parties are creating the project together. This is an important distinction.

THOMAS: There is a tradition of filmmakers working in service of their communities and deeply collaborating with communities as an essential part of the creative process. I am particularly thinking about the history of Black independent cinema in this country. For instance, the L.A. Rebellion filmmaking school [and filmmakers] Julie Dash, Haile Gerima, Billy Woodbury, Sharon Larkin, and Charles Burnett were focused on working within their Los Angeles communities to co-create projects. Fortunately, this has recently been documented in the book *L.A. Rebellion: Creating a New Black Cinema*, edited by Allyson Field, Jan-Christopher Horak, and Jacqueline Najuma Stewart, as well as Zienabu Irene Davis's film *Spirits of Rebellion: Black Cinema from UCLA*. Both of these texts also speak of the influence of UCLA professor Teshome Gabriel and his theory of a Third Cinema—a cinema that is revolutionary in politics as well as in form. Like the L.A. Rebellion filmmakers, Gabriel's work provided me with a blueprint to claim the freedom to sidestep traditional styles of filmmaking as well as feeling the need to conform my work into a specific category such as documentary or experimental. When I look back on my media work of the last three decades, the central focus of the work has been—and continues to be—on co-creation with my various communities, whether it's handing over super-8mm cameras to Black Brazilians in Salvador da Bahia or returning my late stepfather's photo album to South Africa to inspire youth to engage with their history or bringing Americans together to see our community through public sharing [of] our family photographs. As an artist, I claim the space to activate and invite communities on a journey to tell their/our stories.

JUANITA: Yes, I also think that you have to look at the long history of Scribe, which Maori mentioned. Since its founding in 1982, Scribe Video Center in Philadelphia has been engaging communities in media workshops so that the power rests within the communities that we're describing: training community members in camera skills, editing skills, sound recording skills. So, it's not just some random filmmaker coming in to do this work.

This is an important model. Historically I think also we go back to, as you're talking about bridging independent media and television, not only Ellis Hazlip with the public television series *SOUL!* but also people like William Greaves and St. Clair Bourne, who were both with the original *Black Journal* as well as independent documentary filmmakers. We look on the West Coast and the work that the late Lonnie Ding did in San Francisco in really saying that there are voices that have to be heard from the Asian American community. And this is happening in the '60s as well.

I mean, I think about the work of Henry Hampton and Blackside Inc., which in many ways codified the notion of living witnesses as integral to historical documentary storytelling. Blackside's *Eyes on the Prize* also codified the concept of production schools that brought independent filmmakers from different backgrounds together with both scholars and community participants in the planning of the series. Before this, the public television community far too often considered "good history" to be that which was told principally through the voices of scholars, policy makers, and narrators. Acknowledging and giving voice to the people who lived this history as experts in their own right became a whole new way of approaching and changing the narrative that we have to credit Henry Hampton for. So, this is not new, and we can keep going.

MICHÈLE: Another relatively recent example is the Blackout for Human Rights collective, which emerged after the uprisings in Ferguson. Ryan Coogler brought together a collective of artists, activists, and filmmakers across the country to organize artistic events that included music performances, poetry readings, [and] film screenings to highlight the plight of state-sponsored violence against Black bodies in the United States. These events continue to take place annually on Black Friday and MLK Day. Also, it's important to talk about the ways we have co-created and collaborated among ourselves. Filmmaker St. Clair Bourne spearheaded the founding of the Black Documentary Collective [BDC]. The original mission of the BDC was to support each other's work, share information, and in many cases co-create film pieces that it would later take on the Black film festival circuit across the globe and to grassroots screenings. The first five to six years of the BDC were truly active in terms of the creative support we gave each other. Sometimes it's not necessarily about the task of co-creating a particular piece of work but rather creating community and spaces where we can simply support each other's work, whether it is screening rough cuts or other kinds of support needed to keep doing the work collectively and individually.

These initiatives we're discussing and so many others are truly horizontal in terms of approach, bring together people with varying skill sets to express resistance, and are often born out of an urgent political moment that leads to spontaneous collective action. I think that the frustration and feeling of erasure stems from not properly acknowledging the co-creation work happening in our communities for centuries in some way or another.

THOMAS: I just want to reemphasize that filmmakers engaged in these kinds of movements and co-creation processes that we've been discussing have yet to be recognized within academic institutions or even in the field. These filmmakers of color who've prioritized a co-creation framework are often placed in a subgroup within mainstream media.

A case in point is William Greaves, who has foregrounded co-creation as an essential part of his practice. I remember going to a celebration for Bill a few years before his death and speaking to a White Academy Award–winning filmmaker after the event who commented how surprised she was to have never heard of him and wondered why there had been no mention of him in grad school. Yet this same filmmaker was adamantly resistant to the idea that racism could possibly be a cause for this erasure. So, as we talk about recognizing the co-creation work of racialized media makers, past and present, we also have to talk about what is being legitimized, documented, and studied within the canon. We have to make sure that while we're centralizing the histories and work being done in particular communities, we're not marginalizing this work. It's about rewriting and telling a more complete version of the history of this work. For me, that is really important. I observe that my students are hungry for exposure to this work. Even today, it's still so rare that they get an opportunity to see, discuss, reflect, and respond to it.

HOW CAN THIS WORK BE BETTER HONORED AND SUPPORTED?

MARIA: First, these three things: funding our work, exhibiting our work, and writing about our work, such as in film and media criticism and history, [as well as] reviewing it in the journals and trades and news media. It is also important that editors hire and assign writers and reviewers with enough cultural, social, and aesthetic background to capture more of their nuance and context!

Thomas Allen Harris's Digital Diaspora Family Reunion collective (ongoing). Used with permission from Digital Diaspora Family Reunion LLC.

MICHÈLE: I also think it's about having the people of color in positions of power, shaping co-creation conversations and processes. And I'm not sure I am entirely comfortable with this honoring idea. I believe it's even deeper than that. I already know what we're worth, and I know what we've done. Honoring speaks directly to those who have ignored or been unaware of the work that has always been there. Honoring doesn't address deep structural inequity. It's about shifting the current power dynamics, which requires a daily practice of self-awareness by those with power and leverage. There's a deeper dynamic that has to do with the pathology that's outside of us and outside of our communities. I'm not sure how to do that because the problem is not here. Right? So, we're having this conversation, but this is not where the problem is.

THOMAS: But I would say that we are also a part of the solution. We're at a critical moment historically both in terms of our independent work and in terms of what's happening within the industry. We have critical context to add to this conversation. I tell my students it's really important to protest injustice, but it's also important for us to be aware of our power to build as honestly as we can from where we are, which is why the documentation of this conversation is so significant. It's actually similar to another conversation amongst a group of queer African-diasporic filmmakers that Raul Ferrera Balanquet and I organized in the early to mid-'90s, tackling issues around funding and distribution as well as marginalization and invisibility we experienced in the midst of the New Queer Cinema movement. We used our power to write ourselves into history, but instead of this Google document, we used the fax machine to construct the dialogue. This document, entitled *Narrating Our History: A Dialogue among Queer Media Arts of the African Diaspora*, was published in Germany in the mid-'90s and was recently republished in Yvonne Welbon and Alexandra Juhasz's book *Sisters in the Life: A History of Out Lesbian African American Media-Making,* as it is the oldest documented discussion among lesbian media makers. I agree that it's important to identify the problem (whether injustice or inequality), and it's just as important to really interrogate our own power and our relationships with like-minded or like-spirited communities and to be responsible for historicizing ourselves as we make media.

MICHÈLE: I agree with that, and I'm working in universities and exploring the same kinds of ideas. However, there is a work practice of self-awareness that has to happen on the other side. That's what has to happen. We're having this important conversation, and there also needs to be a parallel conversation in White communities about racialized power and where does co-creation practice fit into that.

MARIA: Also, it's not a question of our asking for permission as filmmakers of color, although of course we need funding. What's happening is that audiences are diversifying in the US, and the balance of power in global audiences is changing. As of last year, China overtook North America as the box office superpower, and American media making will become obsolete if our cultural media stories remain provincial and focused on one White, privileged storytelling mode. Audiences want to see a reflection of themselves and are becoming more diverse and international. And in the one media system that is mission-supported, as opposed to profit-supported, in our American media world, I would love to have a more diverse public TV. I would love to have more programmers in that system who are diverse and reflect the complexion of America and who choose to air more diverse media with nuanced and complex representation. I would love more executives in the system who understand that greenlighting and funding diverse programs is imperative to serving the American public given our demographics. Unless we see representation in all aspects of media making, there will be a continued decline in audiences for this media; the field won't be successful.

THOMAS: Yeah. Work has to move beyond simply having a Black face as executive producer or another figurative role which looks good on paper, which is sometimes what is passing as co-creation in mainstream circles right now. There needs to be more investment in communities as well as the next generation of filmmakers that have a deeper level of connection to these communities.

UNPACKING THE COMPLEX ROLES OF MEDIA MAKERS OF COLOR

MICHÈLE: Okay. At every stage of the creative process, I'm asking myself a series of questions like what conversations do I need to have with the editor around character development to avoid stereotypes? How can I create a communal space for assessing and validating the story? How can I be creative in terms of compensating historically marginalized subjects for the labor they are expending in sharing their stories? At every stage of the creative process, I'm asking these kinds of questions and making corrections when needed. This is what is required for disrupting traditional extractive models of storytelling: to be more of a partner with subjects and communities.

THOMAS: Yeah. I just want to say that I watched my stepfather, who was part of the African National Congress, and my biological father, who was unable to be present for his family, both die of alcohol-related disease. And so, healing is a part of my co-creation process. I'm not trying to heal a particular person, but healing in the sense of inviting participants to engage in a healing process that I journey along with them. My projects, like *VINGATE—Families of Value, E Minha Cara/That's My Face, Twelve Disciples of Nelson Mandela*, or *Through a Lens Darkly: Black Photographers and the Emergence of a People*, are all very different formally. But there is a unifying theme in terms of engaging community in the co-creation process. Whether it's a community of LGBT siblings, a South African community where people are not talking across generations, or my current work where I'm asking people to look at each other through our family pictures as opposed to our racial phenotypes, my co-creation work has to do with truth and with healing as a kind of a modality of practice and empowerment through the act of storytelling.

MARIA: It's such an interesting question. For me, co-creation can also be the practice of self-reporting when we're making films about our own communities. Sometimes this is implicit and other times it's explicit because we are so connected to the subjects. And in communities of color, we are likely to have blended families that represent all walks and classes in life, because we know we live in a biased system punitive to communities of color. Yes, today I am a brown US citizen filmmaker and professor, but I have undocumented family, I have family that has been in prison as well as having professionals in my family, including a NASA scientist who sends his experiments to Mars. There are so many stories and worlds that I can explore from inside my community, but sometimes to dare to do that is very challenging and painful and is work that is personal and makes me vulnerable. As filmmakers in our communities, sometimes in the co-creation process to tell an important story we must choose to expose ourselves within the work.

For example, I get so tired of hearing people from outside our communities tell the stories of the undocumented, with so many gaps and errors. Filmmakers from outside the community might say that this is the direct and authentic testimony of this experience because, for example, they "let them speak for themselves." But we know filmmakers cut and edit and frame these stories, and so they may ask the wrong questions or cut the wrong answers. The stories told by outsiders are mediated stories, through an outside lens. I decided to tell my story of growing up undocumented, and I've been surprised when I've been met with resistance. Perhaps it's too raw or too painful or not framed by the familiar master narrative created by outsiders. Perhaps when we use a different frame carved out of our own experiences, it doesn't fit their vision of our truth.

I've been told by funders that they just funded an immigration story, as if the subject had been exhausted or there is only one story and one truth about the immigrant experience. I have had mentors from outside my community tell me that my own mother's character wasn't believable because she did not save her daughter when she was being abused by the stepfather who was the key to their citizenship. That kind of story does not fit conveniently into the simple victimhood framework for immigrants portrayed by some, nor does it fit the story of the criminals and the drug dealers that others might prefer. To many outsiders, "the undocumented" means men standing on corners near Home Depot or the male day laborers on the road to a farm. "The undocumented" has not, until fairly recently, meant the mothers and children living here for years that make up half the eleven million undocumented in the US. Our media focuses on the stories of undocumented children who just arrived and are placed in cages but doesn't want to hear about the millions of undocumented families already living in the US who shake the bars of their virtual cages of systemic oppression. They often lack basic human rights and legal protections as a silent and powerless minority and are an easily exploited workforce for America. As someone who grew up undocumented in America, I see the focus on the attention-grabbing headline, on the horror of children separated from parents and stuck in cages.

To me, there is also a deeper question: what chains must these children rattle for the rest of their lives as long as there is no immigration reform but market conditions encourage this underground exploitation of a labor force? I know that ahead for those children and those families may be even greater horrors of living in a country that criminalizes your very existence, that will not allow you to work, to study in a state university, to have a driver's license, to move freely across borders, to earn a living wage, to seek civil and judicial protections from the many predators that prey on the undocumented. As a person who grew up undocumented, I choose a different frame for these stories. I choose different questions. I may start by looking at the cage, but the longer story of living in that limbo hell for a lifetime feels like such a horror unexplored in the media. That is the story I choose to tell. That's the difference I think between telling community stories from the outside and telling them from the inside. It is that complexity, that refusal to conform your story to the convenient political agendas or boxes that the power structure needs us to fit into. Our authoring our truths is dangerous to the status quo; it is our greatest weapon for breaking the grip of our own oppression.

MICHÈLE: I just want to add one thing to this very important point in terms of these microaggressions that happen in the very pitching of your story: being told your own mother is unbelievable or doesn't exist. Your legitimacy as a filmmaker and a person is questioned, and again this goes back to the idea of power. Who was on the other side making those determinations validating what stories? And so, it leads to all kinds of things around just our own personal health and personal questioning. I'm sure the other side doesn't even realize they've done something wrong.

FOSTERING COMMUNITY POWER AMID SYSTEMIC CHALLENGES

MICHÈLE: I'll use the example of *American Promise*. We would not be able to create that work if it wasn't for the program officers of color who supported us and believed in our story along the way. People like Orlando Bagwell, Rasheed Shabazz, Raquiba Labrie, and Kathy Im. I'm not saying other program officers didn't support the work. They came later. Black program officers came in first and immediately understood our vision in terms of our story and how we worked with our partners as we worked on what our engagement campaign would look like.

JUANITA: I think, historically, the power is in the work that we do behind the scenes that nobody ever sees. Back in the early '80s, as a first-time participant on a CPB [Corporation for Public Broadcasting] cultural affairs funding panel, I was shocked that White panelists were about to turn down proposals by makers of color because they had never heard of this particular artist or issue. And in those instances, you have to sit there and assert that this is the very reason this artist and their story needs to be told. Also, it is important to recognize the work of Black, brown, and Indigenous filmmakers and producers in founding consortia associations to help each other's projects get into the world.

MICHÈLE: Oh yeah. People are like, "Well, I have some money. You got some money. Let's make this move so we can help support this and that project." And I think this kind of cooperation between those with some leverage in the industry is really important to note.

MARIA: I want to jump in here because there's a couple of things. First, I have been mentored by so many and love so many of my Black brothers and sisters in positions of power, and they have been there for me because we have fewer Latinos in those positions as program officers, as funders, as showrunners, as executives in media across the board. And I want to pick up on this concept of co-creation, as our communities of color support each other in funding and creating work. I spent many years as chair of the board of NALIP, the National Association of Latino Independent Producers, now about to celebrate its twentieth year. We're the longest-running organization of Latinx media makers, and many of our programs for the field have resulted in an explosion of Latinx media makers today. And co-creation is absolutely our community of artists picking ourselves up by our bootstraps when sometimes we didn't even have boots, and a number of years ago we began accepting non-Latinx people of color to our NALIP programs because we recognized how kindred we are with access issues around media making.

I started a [residency for women] writer/directors of color . . . across all ethnicities, and I still don't see another residency like it. Right now, it's on hiatus because we lost our very special (and charitably contributed) estate where it was being held. And there's a gap because I have testimonies from women who attended the retreat, many of whom have said it was one of the most formative experiences of their careers, and have gone on to do very well and to lift each other up. I think this was in large part because I also engaged mentors of color from our communities so that we could share collective wisdom of our master artists, and we also called on and honored the value of the knowledge and critiques our fellows could provide for one another. It's valuable to not privilege the guidance coming from people outside of our communities who may be very well intentioned but are lacking the nuance that comes from an understanding of our particular histories and priorities. Often, the microaggressions can happen from those mentoring us from outside our communities, not because they are trying to hurt us—in fact, they are trying to be allies—but they don't always have that necessary knowledge base. That's part of why I and some other artists dreamed that this retreat was necessary—it did not exist before. We had a submission and review process and selected twelve artists for each cohort. It was such a joy to think about the specific needs of our writers and directors of color in a retreat tailored to their needs. As artistic director, I poured so much care into it and so much joy. We had a chef cooking our fellows multicourse, farm-to-table meals so they felt how valued and cared for they were, and had beautiful rooms with fresh flowers everywhere. We had forest and mountain, fire circles and waterfalls, and walks and picnics, and we had so much story and art and supportive criticism and deep elevation of our work, and each of us questioned and pushed and held each other up but also respected and trusted and believed in one another's capacities. I am hoping to run that Artist Retreat Center again if I can find the support!

MICHÈLE: Maria, there's no reason why that retreat that you started shouldn't continue to be funded. The need is clearly there.

THOMAS: Agreed! I think that's the kind of thing that funders should actually be seeking out to fund, especially given where we are right now in the media landscape. I mean that kind of nurturing or workshop or residency that can really lift folks up and also comes from a person-of-color perspective that reorients assumptions around who gets to speak for whom. We are at a place now where we can question what is mainstream and who is mainstream and why. For the television series I am working on now for, PBS, I was mentored by the Center for Asian American Media [CAAM]. I am obviously not Asian American, but CAAM was able to actually see the value of my project and how it contributed to their larger goals around representation and public media. Things are no longer siloed, and we're realizing a kind of win-win narrative as opposed to "I've got to get mine," which is so much of the ethos of this industry on a certain level and can be very destructive. We're supporting each other and giving back.

MICHÈLE: While giving back and mentorship is a great thing to do and [is] important, it's also draining. It's part of what many of us feel we have to do. We are constantly supporting up-and-coming filmmakers of color, but, you know, there's only a certain amount of bandwidth. And I think in some cases if we don't carve out time for our own creativity, our own creativity suffers because that requires head space as well.

The other thing that will lead to this idea of empowerment is more transparency in funding and how money is distributed in the field. Give me what the real numbers are. The Center for Media and Social Impact and the International Documentary Association [IDA] released a study last year on the state of the documentary field based on data gathered from nonfiction filmmakers. I felt there were some key questions around money, power, and race and class that were not directly addressed. And more importantly, we still don't have the numbers from funders and foundations. It's one thing for us to answer a survey and another to find out the lay of the land from foundations, broadcasters, and streaming entities/platforms. I want to know what's being funded and also where do I fit in the larger landscape or field. In order to see that, I need to know from the funding and investment side what their statistics are. If that makes sense? Then we can get an accurate picture and have a deeper understanding of what sustainability looks like and where self-reflection and intervention may need to happen to achieve a greater sense of equity and reimagine the possibilities of co-creation. With an accurate assessment of what the funding landscape looks like, I can make an informed decision about the directions I can pursue in the field. I feel the field is still far from transparent. We don't have the statistics when it comes to those who hold the purse strings and power to greenlight. The studies of those trends need to come to the light. It's something that I'm hoping IDA does a deeper dive into so that we as storytellers can better understand the funding landscape but also so that we as a field can render those with funding power more accountable.

THOMAS: And the idea also of sustainability.

MICHÈLE: Yes.

EXPLORING PERSONAL COSTS AND SELF-CARE WHEN CO-CREATING

THOMAS: I think it's important to take vacations and mental health breaks. That's difficult to schedule in the middle of a production schedule, and you always wonder if others will think you are working hard enough even when you've been working beyond capacity. But it's a question of self-care, even basic things like taking the time to address your health and mental health and scheduling space to recharge the batteries. I say this as an act of affirmation, because it's very hard for me to do.

JUANITA: I've sort of been thinking about what mental health would look like for me, and I have none, I just realized. But the stories that documentary filmmakers tell often get inside of us. On top of that is the issue of consistently having to deal with microaggressions all along the process. I realized that as an executive producer, I oftentimes absorbed a lot of negativity and bias from program gatekeepers [so as] to shield my filmmakers from it so they could concentrate more on creating the work.

MICHÈLE: I think for me, the self-care is building a community of peers who support each other, whether it's being able to vent or create spaces that advance our work. It's having solidarity. In those instances, I can feel the burden lighten. I totally get what you're saying, and I want to include more of that kind of self-care in my life. Just throwing that out there.

MAORI: I am really troubled, I think, personally and professionally by this question because there doesn't—I am not proud that I don't have the space to take vacations. And I am kind of caught in the middle of—I think sort of generationally as a late Gen X person, I have all these millennials who work for me who are all about self-care and all about vacations, and I can't be upset with them at all, right? They say things like "I can't come in today because I'm having low mental health," and they're so clear about it. And the ones who are working, I'm very proud that they have the space to say this. And it isn't like they are making an excuse. But I'm also kind of baffled because I was trained by people who didn't take breaks, and so I'm kind of caught in the middle of that.

And so, I've been burning the candle at both ends for fifteen-plus years of my career. And I'm not really sure how to sort of move forward. I've been making films and doing this festival work on the side of full-time jobs. Sometimes it's been teaching, other times it's been other nonprofit work. All of those jobs have been sixty-hours-a-week jobs, and then taking on another sixty hours somehow. And I know I'm not the only one doing that. I think the field is asking us to do that. The expectations require that you work this hard.

And then, I think this is stating the obvious, but for myself, being a woman and then being a person of color, feeling the need to have everything be in place in a certain way so that it's presented correctly so that there are no questions asked with the budget and proposal. You know, it also requires that you work even harder. No one should work this hard. The idea of self-care is very fraught on a daily basis, and I don't think I'm alone in that.

MICHÈLE: Yeah. Just to bounce off of that, it's so important what you mentioned and this idea of perfect presentation. It's an extra burden of needing to make sure that our work is foolproof so we're taken seriously. Mediocrity is not an option for us.

MARIA: But also, thinking about this one precious life, there are only so many hours in the day, so many years in a life, and how do I serve my community and honor my own expertise and unique and authentic voice and artistic output also as part of that community? What stories do I feel must be told? And I'm an artist, I'm not just a translator, so I want to work with the poetry and language of my medium. And I'm finally at the point where I understand the game and see where the funding trends are, and sometimes I need to say no, that's not what I'm doing, because there is an even greater need I feel to tell a different story or tell this story in a different way than what the trend or the master narrative is demanding. So sometimes, self-care is choosing your own path as an artist and storyteller, even when the larger system tells you that's not how you're going to thrive within that system today.

JUANITA: Maria, you said this so well. Our work is about the authentic voices of our communities, about our own authentic voices as artists, storytellers who are members of communities. I'm so appreciative of the work, experiences, and perspectives that each of us has brought to this discussion. This entire community is so powerful; this conversation has been healing and a form of self-care in itself.

[Agreement and appreciation expressed by all.]

2

NOTHING ABOUT US WITHOUT US: CO-CREATION WITHIN COMMUNITIES

Community-based methods are the most commonly identified forms of media co-creation, both in real life and online. These are projects that put people with first-hand experience at the center of a practice rather than the artistic vision or agenda of a (usually) professional media maker who extracts stories from "subjects" and then displays them to "audiences." Community co-creation makes its watchword "nothing about us without us," from the Latin phrase *nihil de nobis, sine nobis*, an expression that for ancient Roman regimes meant that no policy should be created without the participation of the people who would be affected.

In this chapter we survey the history of this movement toward co-creating with the people formerly known as subjects, participatory art and media that take in the contributions of the people formerly known as audiences, and collaborative forms in emerging technology and journalism as part of the quest to (re)build the commons.

THE PEOPLE FORMERLY KNOWN AS SUBJECTS

Nonfiction filmmakers have often been at the forefront of innovation with emerging technology. More than 90 percent of the films copyrighted in the first decade of cinema were documentaries. Some of the first color films, the first sound films, and the first uses of portable synchronous-sound technologies were documentary. So too, when cameras came off the tripods and documentarians literally took the technology and ran with it, they followed life as it unfolded in front of the moving camera.

Collage created by Helios Design Lab. Photo courtesy of Folk Memory Project.

These highly adaptable forms of innovation are closely connected to extended circuitries of co-creation yet are often attributed to single authors.

As the twentieth century progressed, the documentary field became heavily professionalized. Because equipment was expensive and cumbersome, it encouraged specialization. The documentary field entrenched an inequitable power dynamic between the filmmaker and the film subject, which formed the basis of extractive approaches to documentary. As the documentary scholar Patricia Zimmermann told us, "With the rise of industrial capital is the rise of professionalization, and professionalization means the division of labor. It means skills, not ideas. It means top-down control. It means individualism. I think that in the field of media, the really long, complex, international histories of collective, collaborative, co-creative work have been erased, just erased."

Yet challenges to top-down extractive tendencies in documentary occurred surprisingly early. As Mandy Rose, a British scholar and founder of i-Docs, puts it, "The idea of the documentary subjects' rights has surfaced notably alongside political movements through which those excluded from systems of power have fought for their voice to be heard."[1]

The Scottish-born filmmaker John Grierson, who would become a founder of the National Film Board (NFB) of Canada, coined the term "documentary" for nonfiction film in 1926 and quickly put the word into practice, first as a government film officer and later as a private producer. He began to make top-down, large-scale, industrial documentaries and wartime propaganda. Those early efforts, however, were soon challenged by his sister, Ruby Grierson, who worked as an assistant on the film that became *Housing Problems* (1935), documenting living conditions in London's East End. "In a legendary incident, related by Grierson in his memoirs," Rose writes, "[Ruby] invoked a do-it-yourself ethos, inviting the slum dwellers to tell their stories directly to camera."[2]

Metaphorically handing over the recording equipment, Ruby urged them to take the opportunity to state their case. "The camera is yours. The microphone is yours. Now tell the bastards exactly what it's like to live in the slums."[3] The East Enders' forthright, albeit self-conscious testimonies are still arresting and affecting today, speaking to us across the years.

Documentary practice, especially early ethnographic film, is often associated with anthropological studies and that field's sustained complicity with colonialism and imperialism. Even in socially engaged documentary forms, there remains the problem of the unidirectional gaze and inequitable power relations of maker and subject.

In John Grierson's documentary Housing Problems (1935), residents speak directly to the camera—on the invitation of Grierson's sister, Ruby Grierson—to address politicians and civil servants about abject housing conditions in the slums of London. Image under fair use.

Alta Kahn at her loom. Screenshot from A Navajo Weaver *(1966) filmed by her daughter, Susie Benally, as part of the Navajo Film Themselves series. Image under fair use.*

Community-based co-creation in documentary aims to challenge the gaze of the professional documentary maker(s) and the construction and reproduction of the Other. "The anthropological approach is the dangerous assumption that the observer is superior," said Heather Rae, a Cherokee independent documentary and fiction producer at a Ford Foundation convening at the Sundance Festival in 2018.

In an attempt to address these power inequities, community-based documentarians and artists open up the process of media making to center members of the community and movements across a wide spectrum of co-creation. "You can tell when a story is told from the outside," said Lisa Jackson, an Anishinaabe filmmaker.

Importantly, Tabitha Jackson of the Sundance Institute warns against essentialism in considering notions of insider and outsider perspectives. "We're all outsiders in some way. It's a very blunt instrument in this kind of conversation, and I think for me, it's less about who is telling the stories than who isn't telling the stories. Being an English, mixed-race person shouldn't necessarily preclude me or expect me to only tell the stories of English mixed-race people. For me it's about ethics, the underlying worry of what has been the power dynamic and so therefore what have been the narratives that have shaped our culture. It's absolutely urgent to deal with."

THIRD CINEMA AND THE POWER OF THE COLLECTIVE

There's a long history of collectives of artists, filmmakers, and activists that have organized around the voices of marginalized groups and political movements. These formations often arise in times of political and economic crisis, with rapid responses afforded by the power and anonymity of the collective. Written in the years immediately following the 1917 Russian Revolution, when revolutionary promises were as yet unbroken and a new aesthetic was fast taking form, Dziga Vertov's manifestos have continued to inspire documentary makers. Addressing the societal potentials of film as envisioned in his Kino-Eye project, Vertov wrote that "The textile worker ought to see the worker in a factory making a machine essential to the textile worker. The worker at the machine tool plant ought to see the miner who gives his factory its essential fuel, coal. The coal miner ought to see the peasant who produces the bread essential to him. . . . Workers ought to see one another so that a close, indissoluble bond can be established among them. . . . Kino-Eye pursues precisely this goal of establishing a visual bond between the workers of the whole world."[4]

Perhaps best known for their endeavor "to catch life unawares," Vertov and his Kino-Eye group formed the genesis of the direct cinema and cinéma verité movements that blossomed forty years later. And by that time, the 1960s and 1970s, film/documentary collectives around the world also flourished, inextricably tied to many social and political upheavals as well as technological breakthroughs, eventually including portable video. Third Cinema, a movement first named in Argentina in 1969, decried neocolonialism, echoed Vertov's critiques of capitalism and the institutional modes of storytelling that would come to dominate Hollywood entertainment, and radically challenged the role of the author.

Bolivia's Grupo Ukamau, formed in 1968 by Jorge Sanjinés, Oscar Soria, Antonio Eguino, and Ricardo Rada, among others, made their film *Blood of the Condor* (*Yalwar Mallku*, 1969) with the help of Indigenous villagers who appeared in their recounting of the US Peace Corps' alleged secret sterilization of Quechua women under the guise of providing medical aid. Later Ukamau members became critical of their own stylistic approach, and in their next major film, *Courage of the People* (*El coraje del pueblo*, 1971), took care to construct the film together with the striking mining community

Still from Man with a Movie Camera *by Dziga Vertov (1929). Image under fair use.*

Still from Grupo Ukamau's Blood of the Condor *(1969). Image under fair use.*

that was the subject of the film, since they "had a good deal more right than us to decide how things should be done."[5] In *Problems of Form and Content in Revolutionary Cinema* (1976), Grupo Ukamau wrote that "a film about the people made by an author is not the same as a film made by the people through an author. As the interpreter and translator of the people, such an author becomes their vehicle. When the relations of creation change, so does content and, in a parallel process, form."[6]

In the United States during the 1930s, labor organizations such as International Workers of the World and Workers International Relief and related groups such as The (Workers) Film and Photo League and NYkino explored different types of community relations that would inspire future generations. In 1973, Third World Newsreel grew out of the New York City–based Newsreel in an attempt to strengthen its commitment to developing filmmakers and audiences of historically marginalized communities such as Black and Indigenous communities. The collective was on the ground with the antiwar movement, the Black Panthers, the women's movement, and the American Indian Movement, filming and documenting their histories. Third World Newsreel continues to operate, nourishing the archive, training and promoting emerging filmmakers, and producing film as well as holding community-based workshops tailored for seniors and youths held in public libraries.

With the emergence of portable video gear and the spread of cable television in the 1970s and 1980s came the launch of numerous video collectives.[7] In the context of the United Kingdom's unique Film Workshop Movement, the Black Audio Film Collective was formed in response to the massive economic and political upheavals taking place under Margaret Thatcher's government. According to Helen de Michiel and Patricia Zimmermann, "The Collective's artistic strategies pulled documentary away from its realist moorings into poetic representations critiquing racial imaginaries."[8]

Similar projects appeared and continue to thrive in North America from the hills of Appalachia to the canyons of New York City. These include Appalshop, Downtown Community Television, Paper Tiger Television, and Deep Dish TV. "Deep Dish TV was the first nonprofit organization to distribute programming via satellite [to air on the] local public-access channel," according to Cynthia Lopez, a veteran documentary producer, executive vice president and coexecutive producer of the award-winning PBS *POV* (Point of View) documentary series, and former film commissioner of New York City. In the 1990s when the global justice movement was in full swing, a new wave of autonomous media groups responded. These included Mexico's Chiapas Media Project, Brazil's ISA, Instituto Socioambiental, the Inuit film collective, Arnait, and India's Raqs Media Collective.

"Women in America Earn Only 2/3 of What Men Do" (1985). Photo courtesy of Guerrilla Girls.

Sasha Costanza-Chock, who was part of the Independent Media Network video-production crew in the United States in the 1990s, described it as a profoundly responsive and generative period for media making and movements. Twenty years later, Costanza-Chock continues her work at the intersection of media and movements in collective collaborations but now with the tools of digital culture and design justice. The mantle is also being carried by new generations of media collectives such as the Allied Media Conference and Allied Media Projects, both of which work locally in Detroit and combine social justice work with media making. They hold an annual conference that brings together national and international organizations that are loosely united around a central principle of media-based organizing. As Salome Asega, director of NEW INC. at the New Museum (New York City) told us that "I love these newer collectives who are challenging institutional proximity, always asking their followers, people who engage with them [and] their communities, to just be mindful

A temporary Independent Center, also known as Indymedia, in Edinburgh covering protests at the 2005 G8 summit. Image under © CC BY-SA 3.0 license.

of the ways museums and cultural spaces frame stories. Then they also invite people to produce events with them in said cultural spaces as a way to decolonize them."

One example would be the Chinatown Art Brigade, which co-creates large-scale outdoor projections with pan-Asian tenants' rights, antieviction, and antigentrification organizations in New York City's Chinatown. But the contemporary media-collective formation is a global phenomenon. The Nest Collective in Nairobi was formed in 2012 by a group of twelve artists from diverse disciplines. For their first project, they traveled around the country and collected interviews with over 250 Kenyans who identified as queer. The project resulted in a book and an award-winning film, *Stories of Our Lives*. Njoki Ngumi, a member of the collective, told us that "We take the

Chinatown Art Brigade co-creates large-scale outdoor projections. Used with permission from Louis Chan.

idea of collective authorship very seriously, as a modern remixing of the fading communal approaches and division of labor practices of some of the people of Kenya. We write, produce, shoot, [and] publish all our work ourselves and work hard to eliminate industry-standard hierarchies in both our work and our internal structure."

PARTICIPATORY MEDIA

Even among collective and Third Cinema practitioners, concerns about divisions between specialized artist-observers and their subjects, between professionals and

Prison X (2021) is a virtual reality (VR) piece co-created by a Bolivian team of documentarians, technologists, sound artists, and a fashion designer. "Theatre, Dance, Film, VR, AR, AI, Robotics and other emerging mediums of communication are the tools to tell our stories in all their multiplicity and complexity," said Violeta Ayala, Quechuan director. "We must use their power as a tool for decolonization. Technology is good as long as we control it, but the moment that tech becomes our puppeteer we are lost. The devil is in the detail." Used with permission from United Notions Films.

nonprofessionals, led to further innovations in community-based practices. These are often grouped together as participatory media, which has parallels with terms such as *participatory development* that were also at play in the global South. However, it's a wide category that has often been debated. For our purposes we define participatory media as projects in which the actual tools of production, most often the camera, are directly controlled and operated by community members. This practice accelerated in the 1960s and 1970s, when handheld cameras and then portable video made the technology more affordable and easier to operate. In more recent decades, more widespread access to consumer electronics has made possible a broader array of homemade, community, and online video projects, available on platforms such as YouTube. While participatory media helps redistribute power by sharing resources and modes of production, the point of co-creative contention remains what role, if any, the participants play in the design of a project as well as in the final editing, presentation, and distribution.

One of the most notable early examples was developed in Canada within a thirteen-year-long government-sponsored NFB program called Challenge for Change. Headed by the legendary American documentarian George Stoney, the program experimented with a wide array of co-creative methods, including participatory media, to help historically marginalized communities tell their own stories. In an early Challenge for Change project called *VTR St-Jacques* (1969), directed by Bonnie Sherr Klein, put video cameras and closed-circuit television into the hands of residents of a low-income Montreal neighborhood. The citizens then played back their footage to each other to identify common community issues.

Challenge for Change's best-remembered project was made on Fogo Island, off the shores of Newfoundland. Overseen by producers Colin Low and Donald Snowdon, the Fogo process led to the production of twenty-seven documentaries made not so much for the broader public as to communicate to government officials the islanders' reasons for resisting resettlement plans. The NFB proposal for the project noted that at all stages, the goal was to involve the community in the decisions to be made. The people would select the topics and be involved in editing decisions when the films were played back; they would also determine the extent of the distribution of the films if in fact they decided that the films should be seen by others.[9]

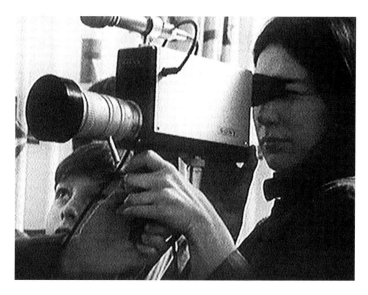

A still from an early NFB of Canada Challenge for Change project called VTR St-Jacques *(1969), which put equipment into the hands of residents of a low-income Montreal neighborhood. Used with permission from the National Film Board of Canada.*

A still featuring fishing boats from the 1967 short film The Founding of the Cooperatives *is one of twenty-seven in the series produced through the Fogo Process of the Challenge for Change program. Used with permission from the National Film Board of Canada.*

As Jeff Webb, professor and Newfoundland historian, describes it, "Individuals and groups discussed their lives before the camera, the film was processed, cut roughly, and [shown] to people on the island to stimulate discussion. In a particularly interesting example, the films were shown to provincial government officials, whose reactions were filmed and that film shown to people in Fogo."[10]

Webb goes on to note that while Low and Snowden's contributions were important to the project and ultimately swayed the Canadian government to keep the island running, scholars tend to neglect the contributions made by individuals of the Fogo Island Improvement Committee and other residents. The formation of the Fogo Cooperative, for example, largely predated the arrival of the NFB crew. Webb's injunction serves as a reminder that the old habits of top-down historiography sometimes impose themselves on processes that were robustly co-creative.

In the United Kingdom, the BBC also experimented with co-creative techniques across nearly thirty years within the Community Programmes Unit, which brought repressed and marginalized communities' points of view to prime-time slots on BBC2 television. The *Open Door and Open Space* series included, for example, *It Ain't Half Racist, Mum*, a critique of TV representation from the Campaign Against Racism in the Media, co-presented by cultural theorist and political activist Stuart Hall in 1979. The unit went on to launch the *Video Diaries* series in 1989, followed by the six-year-long

Video Nation was a BBC television project launched in 1993. Image under fair use.

project *Video Nation,* overseen by producers Chris Mohr and Mandy Rose. They gave camcorders and training to fifty diverse people around the United Kingdom to record aspects of everyday life that they wanted to share with television viewers during the course of a year. These recordings were then edited by the BBC team, with participants seeing and approving the finished product and holding a contractual right of veto. While the work of the Community Programmes Unit was articulated as "Access TV," Mandy Rose has come to think that this term elided the involvement of the media professionals. "I [prefer] the term "co-creation," because it invites you to pay attention to the role of all the parties. . . . It highlights that and therefore allows it to be inter-rogated. It asks that you clarify . . . the terms of the relationship."

In Brazil in 1987, Vincent Carelli initiated Vídeo nas Aldeias (Video in the Villages), an effort that Patricia Zimmermann says "provided video tools to Amazonian peoples belonging to over two hundred distinct groups . . . and not only encouraged self-representation by people often filmed as ethnographic subjects but facilitated intertribal communication as well."[11] The project continues in evolving forms today.

Other current and ongoing projects that place production tools in the hands of citizens include Video Volunteers in India. This project develops a trained cadre of over 240 video correspondents across rural India who create reports and campaigns aimed

Brazil's Vídeo nas Aldeias has been working with Amazonian peoples since 1987. Used with permission from Vincent Carelli, Vídeo nas Aldeias.

at community development. Insight Share, based in the United Kingdom, is another example of a program that does projects across Latin America, Africa, and Asia using "the best aspects of communications technology and participatory techniques" to support communities in devising their own solutions to local issues.

Participatory methods can involve layers of relationships and many forms of art. Since 2010, the *Folk Memory* project in China has invited young filmmakers to participate with the collective by encouraging students visiting their home rural communities to document the historical experiences of relatives and elders during the Great Famine of 1959–1961. This body of work is growing; the collection now includes over 1,400 interviews. The collective's team co-creates with the students, who in turn co-create with their families in the villages. The methodology mixes documentary, oral history, dance, and theater and results in community performances, international tours, and, most importantly, a reclamation of buried personal and collective histories.

Perhaps the most high-profile feature film in the arena of participatory media is the 2004 Oscar-winning documentary *Born into Brothels*, in which British-born photographer and filmmaker Zana Briski and American codirector Ross Kauffman placed

Based in Beijing China, the Folk Memory *documentary project collective invites young filmmakers to visit their homes (in rural communities) to document the historical experiences of relatives and elders during the Great Famine of 1959–1961. Used with permission from Zhang Mengqi, Folk Memory Project.*

still cameras in the hands of children of sex workers living in Kolkata. The film was part of a much larger project managed by the nonprofit organization Kids with Cameras, a broad coalition of organizations and community-centered and participatory interventions. As Lina Srivastava, the organization's executive director at the time the film was in theatrical distribution, wrote in her blog, "The organization's work to scale its photography workshops to other locations inspired a number of other individuals and organizations to put cameras into the hands of children and affected populations around the world."[12]

The film itself, however, focused heavily on Briski and her relationship with the children, at times appearing to center her work in the narrative over and above organizations on the ground. In this framing, handing over a camera can seem problematic, especially if it depicts a humanitarian-development framework that suffers from a lack of accountability and acknowledgments of both formal and informal power structures. While the community-facing work may have tried, the film itself failed to address these questions: How are the images circulated, what forms of agency are activated, and on whose terms? Pooja Rangan interrogates the film in her book *Immediations*, about the humanitarian impulse in participatory models in documentary.[13] As documentarian and scholar Elizabeth Miller suggests, "Emergent documentary forms are encouraging many directors to embrace more participatory and polyvocal modes of production; yet there is still a wide range of interpretation of how much agency or involvement subjects or users have."[14]

BEYOND PARTICIPATION: INTERVENTIONIST MEDIA

Not all community-based, co-creative documentary practices involve only participatory media methods. Many situate their mandate within an interventionist approach. While participatory projects tend to include community members only once the project is designed, interventionist media projects, by our definition, bring participants in as equal partners in designing the project. They create inclusive and open design spaces where joint visioning and decision-making can occur. These projects and programs may match skilled technicians and media or artist professionals with communities and often include sharing tools and developing jointly conceived platforms and campaigns that address community needs and challenge systems of exploitation.

One pioneering interventionist program involved handing over more than a camera; it was an initiative to hand over an entire broadcasting system. In 1970, George Stoney left the NFB's Challenge for Change program and went to New York City to help establish public-access television, simultaneously working on the legal and theoretical framing of the idea for the Federal Communications Commission. This allowed for the general public to create content for television programs and was a prototype of YouTube, perhaps, except that Stoney was not interested in making people famous—he was interested in civic engagement "to celebrate the ordinary things people do to help one another."[15] As Stoney recalled in a 2005 radio interview with *Democracy Now*, "We look on cable as a way of encouraging public action, not just access. It's how people can get information to their neighbors, and their neighbors can get out on the streets to organize."[16]

Subsequently, in 1993 Peter Gabriel and the Lawyers for Human Rights created the New York City–based organization Witness to arm global human rights activists and their communities with the new lightweight digital camera tools of the time. The original intent was participatory in nature, as technically Witness "hands over equipment" but in a different context from participatory media. Witness's methodologies intersect with fields of media, human rights, and technology. Members co-create training programs, grassroots campaigns, and targeted, political advocacy goals with participants. This gives a sharp interventionist focus to projects. The intent of Witness is not so much humanitarian and developmental as it is to collect community-based visual evidence to fight human rights cases in the public sphere, including with governments and the

George Stoney, filmmaker, activist and teacher, 1916–2012. Photo by Philip Pocock. Used with permission from Artists Rights Society, New York.

The Nakamata Tribal Coalition in Mindanao, Philippines, accepts a video camera from the organization Witness in a traditional ritual. From Seeing Is Believing *(2002). Photo courtesy of Katerina Cizek.*

courts. Together with Peter Wintonick, Katerina Cizek codirected an independently financed documentary television film about Witness, *Seeing Is Believing: Human Rights and the News* (2002). Witness has continued to evolve over the years, with thoughtful explorations of the social and political transformations accompanying the changing landscape of media and technology, such as, more recently deepfakes. In its media lab, Witness is exploring the idea of co-presence. This involves live-streaming a distant witness into a dangerous situation in order to have a remote presence, offer advice, and act in solidarity with frontline activists. As Sam Gregory, the program director of Witness, explained:

Witnessing implies the obligation to watch and to act. What is it that we can do that involves bringing people into a live stream that makes them active agents in the narrative as it goes on, driven by the needs of the frontline activists? Not to tell people on the front lines what to do, but to do things that could change events by [either] the support they provide in the moment or generating greater visibility around events as they happen. For example, by translating or adding context or sharing it rapidly. Or can they provide specific skills that might actually influence events as they happen, such as . . . a skill set as a legal observer?

Other organizations connected to people's movements have been grounded in similar principles. B'Tselem, based in Tel Aviv, provides and supports video technologies for Palestinian families and communities as a means for documenting the occupation and violence of the Israeli army. Videre, also a nonprofit organization based in Tel Aviv, provides human rights activists around the world with undercover technologies to record and distribute evidence of corruption, extortion, violence, oppression, and collusion by state and extrajudicial players.

Influenced and inspired by her experiences with Witness, Katerina Cizek sought to apply interventionist approaches when she was invited in 2004 to join the NFB, with a mandate to reinvent Challenge for Change for the digital age. The resulting five-year project, *Filmmaker-in-Residence*, was based at an inner-city Toronto hospital. There, Cizek sought to co-create mostly outside the walls of the hospital in order to explore the potential of health care and media methods in the community rather than within the institution. The project's team jointly designed projects from the outset with community partners. Doctors, nurses, and health care professionals were less interested in holding a camera to frame the image than in framing the concept and intentions of the work, the design of the research, and the ethics of working with vulnerable people as well as acting as accountable advocates and mediators for subjects in the filming process. From the beginning, they co-created and implemented strategies for impact, including successful campaigns to change policy at the hospital level, at both the municipal and provincial government levels, and with the local police board.

A still from the web-documentary, Filmmaker-in-Residence, *in which Katerina Cizek was asked to reinvent the NFB Challenge for Change program for the digital age. The project ran at St. Michael's Hospital for five years, co-created with doctors, nurses, researchers, patients, and community groups. Photo courtesy of National Film Board of Canada.*

An installation view of Rashin Fahandej's A Father's Lullaby *(2017–ongoing), co-created with community members and students in and around Boston, Massachusetts. It is a series of interactive public installations, community engaged workshops, and a geo-located participatory website. Intimate interviews, songs, and lullabies offer poetic meditations on the spaces of love and trauma, presence and absence, and the power of personal memories to interrogate the structural violence of mass incarceration. Photo by Aram Boghosian. Used with permission from Rashin Fahandej.*

The 2021 Oscar-nominated film *Crip Camp* embodies the notion of "nothing about us without us" in its co-created, interventionist process. Codirectors Nicole Newnham and James LeBrecht chronicle the disability rights revolution in the United States from within by drawing on extensive personal archives and recollections of the movement by and of the people directly involved, as LeBrecht himself was a camper at the revolutionary Camp Jened as a teen with spina bifida and living in a wheelchair. The release of the film, coinciding with the global COVID-19 lockdowns, was tied to a powerful fifteen-week series of online digital teach-ins, which aided in addressing continuing accessibility challenges for participants with disabilities. It's fitting that the slogan "nothing about us without us" itself came into usage in English in the 1990s within the global disability rights movement.

RESPONDING TO CURRENT CRISES

Documentary film and art collectives often emerge in dangerous times. According to Okwui Enwezor, a Nigerian artist, theorist, and curator, "If we look back, historically collectives tend to emerge during periods of crisis; in moments of social upheaval and political uncertainty within society. Such crisis often forces reappraisals of conditions of production, reevaluation of the nature of artistic work, and reconfiguration of the position of the artist in relation to economic, social, and political institutions."[17] Some recent examples of this include Mosireen in Egypt and the Kafranbel Media Center in Syria.

Yasmin Elayat, a New-York-based Egyptian-born new media artist, spent several years in Cairo during the 2011 political upheavals. There she joined several art and media collectives that:

all formed literally after the revolution and during the revolution, and all died with the counterrevolution. It was a time when there was a vacuum and leadership [was] completely changing from one hand to the other, [with] curfews—it was not a normal state to be living in in a country. And yet there was the most movement, culturally, artistically; it was almost like a golden era, even though we're talking just about three years. It was some of the most beautiful types of collaborations and art and expressions and initiatives and even organizations [that] were born during this time.

Another strategy from the documentary world emerged in the United States in response to the polarized and volatile political landscape after the 2016 presidential election. While loose communities of documentarians had worked collegially for decades as independent auteurs, they organized in new ways to band together in response to the climate of the Donald Trump administration.

Wonderbox street performances in Cairo, Egypt (2011). Photo courtesy of Yasmin Elayat.

Grace Lee is at the core of a collective of veteran documentarians who have joined forces to document in the two-part film series *And She Could Be Next* (2020) multiple women of historically marginalized communities who successfully ran for various levels of office across the country. Lee has also recently helped found the Asian American Documentary Network, or A-Doc, which is "co-creating a network movement with our peers to uplift and support other filmmakers who are Asian American who are telling stories." Lee stated, "We're very intentional in terms of reaching out to people and looking at the entire landscape."

The Wide Awakes is an art collective that responded to the 2020 elections by devising a twenty-first-century version of an 1860s abolitionist movement of the same name that took to the cobblestone streets of the cities of the Northeast in ritualized nighttime marches, wearing full-length oilcloth capes and black-glazed hats and carrying leaky kerosene lanterns. The modern-day activists re-created those rites in a series of freeform happenings over the summer of 2020 across Brooklyn, Washington, DC, Harlem, Chicago, and Berlin, Germany.

Meanwhile, Firelight Media is a film production and training company that has historically built close relationships with movements, often making connections and contacts at the rough-cut stage of a project. But now, the group has begun a co-creative process on a project before the medium has even been defined and convened with over thirty social change movement leaders from Black Lives Matter, #MeToo, the National Women Farm Workers' Alliance, the Rainforest Alliance, Dream Defenders, and the New Florida Majority. According to then Firelight vice president Loira Limbal in our interview, "We are trying to bring together a diverse as possible cross-segment of communities that have been historically targeted and marginalized . . . to create [narratives] that really contemplate the needs of all of us. Kind of like, all of us or none of us. How could we co-create a movement media alongside our organizational partners? What does that mean? What are the best practices that we need to be thinking about? How does that get funded?"

Many of these movements call for equity in media representation, production, and representation with a foundational acknowledgment and accountability of historical wrongs. But for some, there is more to it than equity.

INDIGENOUS NARRATIVE SOVEREIGNTY

A new framework, known as Indigenous narrative sovereignty, is being articulated by Indigenous artists and media makers around the world. This framework centers not only Indigenous stories but also the teams of decision makers that make and own the work. Jesse Wente is the director of the newly created Indigenous Screen Office, an independent national advocacy and funding organization in Canada. In a keynote at the Hot Docs Film Festival, Wente said, "Indigenous people are not seeking equity. We are seeking sovereignty." In the Canadian context, this refers to a long-standing but burgeoning political movement related to land rights, treaties, and self-governance. But it also extends to sovereignty over Indigenous narratives.

While the Indigenous Screen Office interacts with industry-recognized production models, it foregrounds the importance of co-creating from within communities, respecting Indigenous protocols, collective authorities, and processes within Indigenous nations. As Anishinaabe filmmaker Lisa Jackson described the position of creators such as herself:

We're always implicated, which is another word that I am using a lot. I think it's important that we feel implicated in what we do, and that we look at our relationships with our subjects in a long-term fashion, and we look at what it brings to them. . . . Especially because in the wake of the [2015] Truth and Reconciliation Commission [of Canada],[18] there's been a huge

A still from filmmaker Lisa Jackson (Anishinaabe)'s Biidaaban: First Light *(2018) VR in which she explores "how the original languages of this land can provide a framework for understanding our place in a reconciled version of Canada's largest urban environment." Courtesy of Mathew Borrett.*

interest in Indigenous stories, which is good on the one hand, but also challenging on the other hand. Because a lot of people feel like they have gotten a sense of the trauma of Indigenous people for the first time. They really feel the need to make sure everybody knows about it. Who does that benefit? Not the Native [peoples]. . . . We all know what is going on there. People are more interested in solutions, also strengths, also a way forward, also honoring historical wrongs like treaties.

For Indigenous peoples, control over narrative is inextricably linked to control over the historical narrative, governance, land, and the future. In its multiple and diverse manifestations, the power of collectives embodies this core principle of co-creation.

IN CONVERSATION: "IF YOU'RE NOT AT THE TABLE, YOU'RE ON THE MENU"

by DETROIT NARRATIVE AGENCY

The stories told about a place form a kind of DNA, shaping what that place is and what it can become. For too long, the stories that circulate about Detroit have defined it as broken, violent, and in need of saving from itself. More recently and especially since Detroit's emergence from bankruptcy in 2014, there's been a new common strand of stories about the city, particularly stories of resurgence led by White billionaires,

The Detroit Narrative Agency's cohort (2017–2018) seeks to shift stories currently being told in and of Detroit as a means to advance social justice. They produced two narrative and three documentary films. Back row from left to right: Natasha Tamate Weiss, Atieno Nyar Kasagam, Orlando Ford, Alicia Diaz, and ill Weaver. Front row left to right: paige watkins, Bree Gant, Cierra Burks, and Ahya Simone. Photo by Kashira Dowridge. Courtesy of Co-Creation Studio.

scrappy entrepreneurs, and pioneering artists. Invisible from that narrative is the Detroit that was saving itself all along, the Detroit that is pushing back against marginalization and erasure, the Detroit that has a vision for a future based in liberation and justice. Black and Brown Detroiters have not been at the table when it comes to narratives around the city's so-called rebirth. And as goes the organizing lesson that Detroit water warrior Charity Hicks helped popularize, "If you're not at the table, you're on the menu."

The Detroit Narrative Agency works to amplify that Detroit by incubating compelling films that shift the dominant narratives about this place. The organization's fellowship program supports a yearly cohort of filmmakers of color to develop short films accompanied by community impact strategies. The Detroit Narrative Agency co-creates media from within communities, and the co-creation happens in multiple layers, in concentric circles. The process starts with deep listening rather than preset agendas.

"Our co-creation is two-fold," says Alicia Diaz, a filmmaker in the Detroit Narrative Agency's first cohort. "It's the community work that we're doing on our projects, but I also think about the community that we have created as all being a part of this cohort. I never expected that. . . . We're also laying a foundation for those who come after us."

The Detroit Narrative Agency's strategies come out of an immense historical legacy of activism and cultural production in Detroit. This includes two decades of what Allied Media Projects defines as "media-based organizing" and over a decade of community building and network cultivation. This context is important; these strategies and tools were developed out of a long-term commitment to radical organizing in Detroit.

Allied Media Projects, Detroit Narrative Agency staff, and Detroit Narrative Agency fellows gathered around a microphone in 2018 to reflect on their process, their origins, and how the work was unfolding. The following is an edited selection from those conversations.

PAIGE WATKINS, ASSOCIATE DIRECTOR, DETROIT NARRATIVE AGENCY: What is media-based organizing?

MORGAN WILLIS, PROGRAM DIRECTOR, ALLIED MEDIA CONFERENCE: We define media-based organizing as any collaborative process that uses media, art, or technology to address the roots of problems and advances holistic solutions towards a more just and creative world. . . . The reality is that people are doing media-based organizing in so many different ways in their own lives, in their own work, and in their communities that they just might not know that term. Or, that term might not be the label that they apply to their work. . . . In some ways I think the definition is aspirational and the essential traits are aspirational, but it's always important to have a clear aspiration towards those things.

JENNY LEE, EXECUTIVE DIRECTOR, ALLIED MEDIA PROJECTS: Yeah, and I wouldn't even go so far as to say there is an ideal that embodies all of these things because what's interesting is the way that projects that come into our orbit . . . resonate with different parts of this definition but also bring new approaches and things that we're not already thinking about.

ILL WEAVER, FOUNDING DIRECTOR, DETROIT NARRATIVE AGENCY: [The Detroit Narrative Agency] exists because it comes out of a legacy of media-based organizing that we have all participated in, in a multitude of ways—through Detroit Summer, through the Allied Media Conference, and through many other realms connected to the network that we're a part of. . . . So, that could be through my work with Complex Movements and Emergence Media; [Morgan], your work with Brooklyn Boihood; [paige], your work with Black Bottom Archives; [Jenny], your hand in a million different projects that you enact these values through. . . . We were really maybe fifteen years into doing this work by the time that Ford Just Films approached us, when Cara [Mertes] came up to us and was like, you know, I really wanna figure out how to work with y'all to support moving image–based, narrative-shifting work. And then, based on that, I feel like there was some co-creation of what that could look like.

JENNY: It wasn't like, "Hey community, here's this opportunity, jump on." It was more like "hey, we have these resources, what should the priorities be? What should the process look like?" You know, this very deep, beginning by listening, back-and-forth deliberative thing. And now it's also translating into the process that you all are supporting the filmmakers through, where they're reaching out or connecting with community organizations working on their issues and not just coming with a singular artistic vision. So, I think that spirit of collaboration is layered into the whole process from the beginning.

The first step for the Detroit Narrative Agency was to create an advisory team that facilitated a process for over two hundred Detroiters to help shape the narrative-shifting priorities of the Detroit Narrative Agency. The advisory team also conducted an audit of existing moving-image media about Detroit. They were able to identify inequitable patterns, such as confirming a hunch that the majority of films made about Detroit—an 85 percent Black city—were being created by White filmmakers from outside Detroit.

PAIGE: We collected a lot of feedback. What do the people of Detroit want to see less of, and what do they want to see more? That was a six-month process.

ILL: Members of the community advisory team then served as a panel to select ten projects. Those projects went through the first year of our programming. It included a seed grant of $6,000 support for each project, access to equipment, mentorship, training workshops, support to create a work sample and a pitch deck. They presented at the Allied Media Conference that year. We also took several folks on a trip to BlackStar Film Festival.

From there, we moved into this new phase, an incubator or fellowship program. Five of the ten projects are now creating a short-film version of their project. This phase also includes developing an impact strategy: developing community partners that . . . work with them as they create the project and, once it's distributed, to make sure that it leverages a larger collaborative constellation of people who are shifting a similar narrative to their project.

*   *   *

paige watkins of Detroit Narrative Agency (left), and paige watkins together with ill weaver (right) (2018). Photos by Kashira Dowridge. Courtesy of Co-Creation Studio.

The five projects in Detroit Narrative Agency's fellowship include two fiction films and three documentaries.

Director Alicia Diaz: "Our co-creation is twofold. It's the community work that we're doing on our [own film] projects, but I also think about the community that we have created as all of us being a part of this cohort." Photo by Kashira Dowridge. Courtesy of Co-Creation Studio.

DANGEROUS TIMES: REBELLIOUS RESPONSES

Director Alicia Diaz has made a documentary and exhibition called *Dangerous Times: Rebellious Responses*. The sanctuary movement of the 1980s was a religious and political campaign to provide safe haven for Central American refugees fleeing civil conflict, a move driven by over five hundred US congregations across eleven denominations. *Dangerous Times* traces the movement's rise in Detroit through the personal accounts of Esther Gálvez, a Latinx sanctuary advocate, and Sihanouk Mariona, whose family was among the most visible Salvadorian exiles in the United States.

ALICIA DIAZ: Being Afro-Latina, and being othered, and understanding the life of being othered and also marginalized—this is helping me to give voice and give a little bit more insight that's unknown in the greater part of the city, especially when you're talking about our history and our immigrant past and present. It's almost like a ghost whisperer, but no one's dead. . . . My whole experience, even though I am American, I'm not an immigrant . . . I still live like that, I still live within that community. And being Afro-Latina it's like I'm in two places at once. And through this film we're able to pull it out.

Director Ahya Simone (right, with co-creator/actor Cierra Burks, left): "I'm a part of this community, so I've never had to think about it. . . . Honestly, the creation of Femme Queen Chronicles *came from just co-creation between a bunch of trans women in a car going to Wendy's." Photo by Kashira Dowridge. Courtesy of Co-Creation Studio.*

FEMME QUEEN CHRONICLES

Director Ahya Simone has made a web series called *Femme Queen Chronicles*. Chanel and her friends Eryka, Amirah, and Shevon all are just trying to make it through the day without getting clocked as trans women—or clocking someone else over the head on the way. *Femme Queen Chronicles* is about the lives of four Black trans women as they navigate love, life, trade, and shade in the city of Detroit. The series is written, directed, and brought to life by Black trans women themselves.

AHYA SIMONE: I'm expanding what it means to be a Detroiter. When you think about Detroit, you don't think about queer and trans people of Detroit. I think *Femme Queen Chronicles* kind of brings queerness and transness to the forefront in people's cultural associations of Detroit so that we can imagine it as more than just the Motor City and Motown and cis-hetero people, and we can imagine the multitude of the types of folks that live here. I think it's shifting narratives about trans people in media, I think in a more humanizing and vulnerable and humorous way. By hiring an all trans cast and being a Black trans woman and writing and directing and telling the story and also having interns who are trans on set and in front of the screen and behind the screen, it's something that's impactful for trans communities of color and trans women of color: to create more creative opportunities for the girls in Detroit.

Director Atieno Nyar Kasagam: "The whole concept of 'working with community' for me, that language alone comes from a colonial place . . . we feel that we are the community. . . . We are locating the land in a very significant place in the film, giving agency and voice and treating land as intelligent, high conscious plants, our ancestors." Photo by Kashira Dowridge. Courtesy of Co-Creation Studio.

SIDELOTS

Director Atieno Nyar Kasagam has made a documentary called *Sidelots*, a love story of Black land reclamation told in ritual between Detroit, Alabama, and Kenya. The documentary follows one family on Detroit's East Side as their story of urban farming unfolds into a spiritual journey of discovery, loss and reindigenization. By digging up familial and land roots across the diaspora, *Sidelots* illuminates all that is sacred in the land and encourages a radical reconsideration of how we view the earth immediately below our feet.

ATIENO NYAR KASAGAM: We are shifting the narrative of vacancy. We are inviting people to see the abundance and the aliveness of side lots in the city of Detroit. We are inviting people to look at over twenty square miles of liberated land as fodder for the redesign, designing of a Black city, a just and equitable city that remembers that it is on great Indigenous land by the Great Lakes, that this land is alive and beautiful and we do not need to grow grass, and we need to end the grass supremacy in the culture. And, jokes aside, inviting people . . . to know that we have what we need. We have the most important resources that we need to take back power and freedom and agency.

Director Orlando Ford: "I'm trying to meet the needs of the community. . . . Seeing my neighborhood and the neighborhoods around me changing and being in disarray and disrepair, it affects me. It's what I see, so my connection is I live here too, and I care about it." Photo by Kashira Dowridge. Courtesy of Co-Creation Studio.

TAKE ME HOME

Director Orlando Ford has made a documentary titled *Take Me Home*. A home fore-closure crisis has gripped Detroit for over a decade. In this time, illegally inflated property taxes have caused more than 100,000 working families to lose their homes. While headlines read of the so-called rebirth of Motor City, many Detroit neighbor-hoods have been devastated, with African American communities hit hardest of all. *Take Me Home* follows one family as they fight to save their home and struggle to keep their neighborhoods and communities from being lost.

ORLANDO FORD: It's just ordinary people going through extraordinary things . . . I've been dovetailing back to the organization and saying, "okay, this is what I've done so far, what else can I do to help this process along," because I'm coming in at the tail end of this, and some of these people have been doing this kind of work for years.

Director Bree Gant: "With this project, I'm co-creating a lot . . . I'm a very mad scientist–type artist and used to creating in my own room. So, the first step I did was just take my headphones out when I'm on the bus and talk to people and to build actual relationships. . . . Showing up, building relationships. There are answers there." Photo by Kashira Dowridge. Courtesy of Co-Creation Studio.

RIDING WITH AUNT D. DOT

Director Bree Gant's experimental fiction film and multimedia project *Riding with Aunt D. Dot* brings together personal narrative, radical imagination, experimental video, and the Detroit city bus. The film tells the story of a disillusioned Detroit artist struggling to ground herself in reality or in her dreams.

BREE GANT: I think one of the most valuable things I've learned is that there's just so much culture on the bus . . . street culture and street fashion, especially. Noting how many trends—straight from the street—go to the runway. There isn't that much room for even claiming inspiration because there are so many things that just get blatantly repeated on the runway and claimed as originality, when really it's just that we aren't aware when these people with power have excavated these cultural traditions from us. . . . I really want the resources and the money that's coming into the city to reach the bus, but I hope that gentrification never reaches the bus because there's just so much culture and originality there.[19]

PHOTO ESSAY: DETROIT NARRATIVE AGENCY

The following collection of photographs was taken during two Detroit Narrative Agency/MIT workshops held in Detroit at the Boggs Center, home of the legendary Detroit activists and writers Jimmy Boggs and Grace Lee Boggs.

Detroit photographer Kashira Dowbridge. Photos by Katerina Cizek. Courtesy of Co-Creation Studio.

The Boggs Center continues to serve as an active, vibrant space for community organizing. The main floor remains as it was during Grace's last days here in 2015. She lived to be one hundred years and one hundred days old. Photo by Kashira Dowbridge. Courtesy of Co-Creation Studio.

Orlando Ford: "It didn't hit me until after I got home of how profound it was just to be in that space and say 'Oh, okay, their house is turned into a center for the community as they lived their lives.'" Photo by Kashira Dowbridge. Courtesy of Co-Creation Studio.

Atieno Nyar Kasagam: "To be in that space was to be truly in the house of the greatest superheroes that you know . . . it's like going to visit Prince's house . . . And I cannot eulogize Grace and Baba James Boggs because I'm too young in the physical body form and I'm also too young into the knowledge of Detroit and the kinds of revolutionary work that they have done, but I know I am in the right space." Photo by Kashira Dowbridge. Courtesy of Co-Creation Studio.

Alicia Diaz: "The kitchen is significant because the kitchen is the heartbeat for communities of color. Yeah, it's the heartbeat, it's where all the stories are told, it's where traditions are passed down. With Esther Galvez, that's where she received the knock on the door, to her from becoming ordinary to extraordinary, into the sanctuary movement." Photo by Kashira Dowbridge. Courtesy of Co-Creation Studio.

Atieno (left) and producer Natasha Tamate Weiss (right). Photo by Kashira Dowbridge. Courtesy of Co-Creation Studio.

Natasha Tamate Weiss: "Grace being what is called an Asian woman, it makes space . . . I feel like she made it possible for me to be here." Photo by Kashira Dowbridge. Courtesy of Co-Creation Studio.

While in the Boggses' library, the Sidelots team discovered two books of personal significance. Atieno found a book by "one of Kenya's most powerful independence leaders called Mbiyu Koinange, and this man was just like me. He studied in the US and in England at around my age," she said. "He was no more than ten years older than I am right now when he wrote that book that was published in Detroit. And I was, like, 'God damn, how did he get to Detroit?' And not only was he in Detroit but [also] the Boggses had five copies, six copies, of the same book." Photo by Kashira Dowbridge. Courtesy of Co-Creation Studio.

Meanwhile, on a shelf close by, Natasha found James Boggs's American Revolutionary *translated into Japanese. "I'm just releasing, surrendering to the permission that has been granted and just really being humbled by that and honored by that," said Natasha. Photo by Kashira Dowbridge. Courtesy of Co-Creation Studio.*

ill Weaver. Photo by Kashira Dowbridge. Courtesy of Co-Creation Studio.

Alicia. Photo by Kashira Dowbridge. Courtesy of Co-Creation Studio.

Natasha and Atieno. Photo by Kashira Dowbridge. Courtesy of Co-Creation Studio.

The Boggses' library. Photo by Kashira Dowbridge. Courtesy of Co-Creation Studio.

Orlando (left) and Atieno with her child (right). Photo by Kashira Dowbridge. Courtesy of Co-Creation Studio.

ONLINE: PEOPLE FORMERLY KNOWN AS AUDIENCES

In earlier forms of documentary, the term *community* usually referred to people who inhabited a specific physical place or shared values and identities or both. In the digital context, the word also refers to online viewers, users, and participants. When documentarians began creating on the web as a native platform—as opposed to using the web to market linear films—there was a radical shift in the potential for the audience to take on agency as co-creators. New implications for documentary continue to emerge in the fast-evolving expansion of the metaverse.

Web-based interactive documentaries began flourishing in the early 2000s—prior to the mass onslaught of social media—at a time when web technology became robust enough to enable video formats more easily. (Cizek's NFB *Filmmaker-in-Residence* is considered the first feature-length online documentary.) Often, these projects merged in-person, community-based work with the nascent tools of decentralized networks of engagement over the internet. Makers of these projects relied on a

A scene from the immersive theater piece £¥€$ (LIES), in which audience members play a form of poker. The Belgian theater and performance collective Ontroerend Goed greet audiences as co-creators. Photo by Thomas Dhaenens and Michiel Devijver. Used with permission from Ontroerend Goed.

cohort of independently built platforms in order to organize community-generated stories. These platforms included GroupStream, Vojo, Cowbird, Mozilla's Popcorn, Storyplanet, Korsakow, Klynt, and Zeega as well as video platforms such as YouTube, Vimeo, and others that were just emerging.

The work of the audience online often happens in navigating the menu of a closed-system online documentary, thereby co-creating a textual path through a myriad of choices. Online documentaries introduce nonlinear, iterative relationships that afford a certain agency to the audience now scrolling through or immersed in a storyworld within which they can maneuver and to which, sometimes, they might contribute. This allows for expanded perspectives and readings. In many cases, these online documentaries also open up space for participation in the form of crowdsourcing and, in the terms of the period, user-generated input.

For *18 Days in Egypt* (2011), Jigar Mehta and Yasmin Elayat experimented with the GroupStream platform as an online space in which Egyptians could share their stories of the 2011 political uprising in Egypt. One of the project's biggest issues was the

The Are You Happy? *project (2010) prompts collaborators around the world to ask people in their communities "Are you happy?" Image under fair use.*

recruitment of participants, an issue facing this entire generation of early user-generated platforms. "A lot of the challenges that we had were the barriers to get people to actually participate," said Jigar Mehta, who now works at AJ+, Al Jazeera Media Network's online news and current events channel. "It turns out it's a lot harder to do that than we thought," Mehta said. "It requires a lot of the things that we're actually all familiar with as storytellers: actually spending time with the people who are sharing their stories, being on the ground, creating that intimacy between the audience and the subject."

Media scholar Henry Jenkins calls this the "participation gap." As Mandy Rose explains the challenge, "Uneven participation may be less to do with the lack of access to equipment or technologies sometimes referred to as the 'digital divide,' and more to do with a lack of confidence in and understanding of the protocols of engagement."

For *Sandy Storyline*, a project about the personal and political narratives emerging from the Hurricane that damaged parts of New York City in 2012, the co-creators also needed to address the issue of recruitment. Michael Premo described it to us as a system of "pathways to participation" involving intense in-person meetings as well as online collaboration. "We had regular open meetings weekly, where anyone could come and was invited to give feedback and input around this core design question: 'How do we collaborate with millions of people to tell a story?'"

The landmark installation and web documentary, *Question Bridge* (2012), also interwove in-person and online methodologies by prompting Black male participants, of diverse backgrounds, to ask and answer questions of each other through the mediation of the project's artists and their cameras.

Yet accessibility issues still exist. In the *Quipu Project*, María Ignacia Court, Rosemarie Lerner, and Sebastián Melo of Chaka Films set out to support the Indigenous women who had endured state-run forced sterilization in the 1990s under the government of President Alberto Fujimori. The co-creators needed to work with sensitivity and thoughtfulness in rural areas of Peru where there was limited or no

Youth watching Question Bridge *(2012). Used with permission from Kamal Sinclair.*

internet access, specifically in remote Indigenous communities. It was also necessary to protect participants' identities.

Women were invited to participate anonymously by using telephones to call in and record their stories as voice messages that could then be organized and distributed through a telephone interface as well as online to gather support from global audiences. The co-creators left phones with women in remote communities and then promoted the project widely via radio broadcasting. As the documentary team traveled through communities, they would also invite one of the workshop participants to come along with them to the next event; this connected leaders of the network through in-person meetings and strengthened the movement.

Importantly, however, when considering interactivity that relies on the web, the internet itself can be misunderstood according to Caroline Sinders, an artist from New Orleans who researches the systems that regulate and proliferate hateful online comments. She argues that we need to understand that the internet is in fact a large system iteratively re-created by many sets of human hands. She brings up content moderation, for example: "How much of that, like how much of the backbone of

A participant in Peru engaging with the Quipu Project (2015) on a mobile phone. Used with permission from María Ignacia Court and Rosemarie Lerner.

the internet, is still maintained by people? So many open-source projects are built by volunteers. . . . I think we think of the internet as this purely digital space when in fact . . . it is so full of people. And I think that maybe a bit of the future is recognizing the humanity of who builds and maintains the internet."

While online networks have connected the world to decentralized and open-source methods, the above projects demonstrate that in-person co-creation is often a crucial component for meaningfully challenging issues of inequity and injustice.

THE RISE AND SWELL OF SOCIAL MEDIA AND BIG TECH

Today, it's hard to imagine a time of making media online before YouTube, Facebook, Amazon, and streaming services. YouTube was bought by Google only in 2006, the same year Facebook became available to anyone with an email address. Amazon turned its first (relatively small) profit in 2001, at a time when Netflix was still mailing its customers DVDs through the postal service, and did so throughout the decade until it introduced streaming as a service in 2010. Yet the aforementioned early independent documentary experiments in online co-creation that emerged then became harbingers for the social media platforms of the 2000s and onward. These are all platforms that make claims about co-creating with their users in one form or another.

Within a mere two decades, an overwhelming takeover by big tech and social media platforms has dramatically narrowed the gateways through which users access the web, originally co-created on the principles of an open and free system to share information. Millions of users in Nigeria and Indonesia, for example, responded to surveys in 2015 telling researchers that while they were on Facebook, they did not believe they were on the internet.[20] An entire generation has now grown up into a world in which social platforms *are* the internet. But these platforms have proliferated with no transparency or accountability. They've been built on business models that financially reward systems that create and maintain addictive behaviors and draw more clicks from polarizing and extreme content. In turn, we've seen these systems embolden the malicious participation of online communities such as QAnon and various right-wing terrorist groups all around the world who have organized and fomented hatred online that led to physical violence in, among other situations, the Indian pogroms in February 2020,[21] the ongoing genocide in Myanmar,[22] and the US domestic terror attack on Capitol Hill on January 6, 2021.[23]

As of January 2021, 4.2 billion people, or half of the world's population, were online.[24] But what does this mean? The internet is not the democratizing force that early open-source inventors and experimenters hoped for and promised. In her book *The Age of Surveillance Capitalism* (2019), Shoshana Zubboff describes in excruciating detail a global technological architecture that creates consumer addicts and tracks, predicts and controls behaviors to turn untold profits for a few. A growing body of scholarship—primarily by African American women—exposes new forms of age-old racism and White supremacy in algorithmic systems, including social media platforms, big data, artificial intelligence (AI), and surveillance systems that all function in the service of police, military, state, and corporate interests to segregate, discriminate, and dominate along lines of race, class, gender, and caste. In *Algorithms of Oppression* (2018), Safiya Noble interrogates how search engines reinforce racism. Ruha Benjamin's *Race after Technology* (2019) charts the rise of a "New Jim Code," a totalizing matrix of oppression. Joy Boulamwini's work, as depicted in the documentary film *Coded Bias* (2020), reveals the biases in facial-recognition software and charts her Algorithmic Justice League's efforts to ban the technology in the United States.

Critics are only in the nascent stages of understanding the enormity of the crisis of the intersecting forces of social media and authoritarianism, as Robert J. Deibert argues in *Reset* (2020). He equates the current moment to the early stages of the environmental movement in the 1970s and 1980s. The growing body of commentary, he writes, has "dissected what's wrong and has helped wake us up to a serious pathology, but [we] have yet to carve out a confident alternative way to organize ourselves."[25]

NEW TECH, NEW DEVICES

Emerging tech available for media making is moving beyond the tethered desktop and even mobile devices onto new devices and untethered spatial networks. In 2015, a second wave of VR projects took the Sundance Festival by storm. Since then, the avant-garde of experimental documentary and the arts have quickly shifted from the interactivity of the web to a new horizon of immersive tech, much of which is tied to hardware that can be inaccessible, complicated, and expensive: 3D printers, VR, augmented reality, and AI. These technologies carry the promise of new emotional, experiential, and interconnected realms and will aggregate, organize, and interpret massive amounts of data on a scale not seen before, especially as access to the metaverse becomes more ubiquitous.

This moment may be evocative of early cinema. When the nascent form of film was still widely understood through the old lenses of theater and literature, it took the radical approaches of collectives such as Vertov's Kino-Eye to invent a new cinematic language. Likewise, the maturation of a language in VR might require a radical

VR Assent, *by Oscar Raby, at Sundance (2015). Used with permission from Oscar Raby.*

VR Project Syria *at Sundance (2015). Used with permission from Nonny De La Pena, Emblematic Group.*

shift from cinematic approaches to new conceptualizations of spatial experiences and explorations. Rather than framing and directing the viewer's gaze in a manner familiar from cinema, VR offers an opportunity to create spaces of narrative possibility that can be explored and rendered coherent by the participant. In this scenario, world designers and participants work together to actualize individual story strands in a relationship that differs fundamentally from the unilateral storytelling dynamic of film, in which the same story is told to many people. The model of discovery rather than direction has significant implications for the work of co-creation.

VR production can be slow and cumbersome, and participatory projects in the field are still rare. But in Brookline, Massachusetts, a permanent new community VR center called the Public VR Lab has launched a space that, according to its website, "values and promotes digital inclusivity, accessibility, training, equipment and XR [extended reality] content in the public interest." This and similar endeavors have

Participants at the Public VR Lab, which is building a new field for community extended reality through accessibility and digital inclusion, best practices, low-cost training, equipment, and co-created traditional/ emerging media content designed to facilitate community dialogue, media literacy, and VR/AR education in the public interest. Used with permission from Ann K. Bisbee.

sparked an array of community-based projects, a number of which have proven to be quite innovative. In the Brownsville neighborhood of Brooklyn, the exciting partici-patory VR game experiment *Fireflies: A Brownsville Story*, launched in 2020.

Fireflies was in the works for three years as a co-creation between the Browns-ville Community Justice Center, Peoples' Culture, more than thirty youths, and over one hundred community members, all from Brownsville. Through workshops and conversations, the team decided to make a docugame rather than a traditional documentary or narrative film. They then created the Tech Lab, which is now a permanent fixture within the Brownsville Community Justice Center. They began with a process-focused lab with no predetermined outcome rather than a product-driven project. "Our main goal, our chief goal, is that the project is a catapult to something else and that the work of art itself lives in any person who is a part of it," said Nicholas Pilarski, co-creator. The training and resulting production involved building custom computers in addition to camera training, unity game design, and mapping open data via a geographic information system, with which they re-created Brownsville within the virtual space of a game that users can travel through and inhabit.

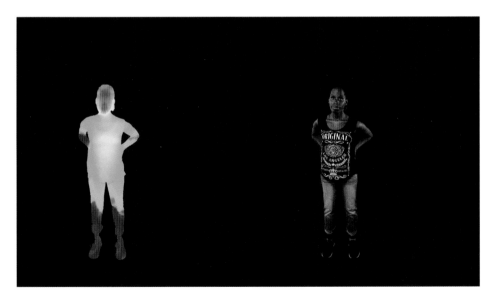

A production still from Fireflies: A Brownsville Story *(2020). Photo courtesy of Brownsville Community Justice Center and Peoples Culture.*

So far, however, immersive technology is cumbersome. Before the resulting images can be seen, the data requires long computer processing times, stitching, and effects building. The technology is in constant flux and is technically complex in ways that can be fetishized. The studio setup for these emergent media projects can also be intimidating and overwhelming for nonprofessionals, particularly those living in or coming from vulnerable contexts. Because of this, projects require the initiators to finesse the weaving of production logistics with ethics. One of the few accessible immersive platforms to emerge so far is REACH, a joint project between DepthKit (Scatter) and the Emblematic Group.[26] In 2019, REACH introduced a simple drag-and-drop interface for users to build worlds inside VR.

One project has gone quite a distance in employing emergent technologies and intentionally considering the implications of a community-based approach. Using holograms and robust AI systems, the Shoah Foundation is working with a community of Holocaust survivors to create *Dimensions in Testimony*. This project allows users to ask holograms of Holocaust survivors questions about their life stories. Using

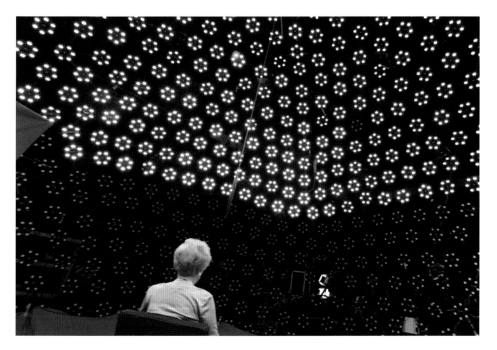

A production still in a 116-camera dome in Dimensions in Testimony *(2014). Used with permission from the USC Shoah Foundation.*

natural-language processing, the system calls up answers that have been developed through the processing of hundreds of hours of prerecordings.

Stephen Smith, executive director of the University of Southern California Shoah Foundation, described for us how the team captured the vast amount of data required for the project. They placed Holocaust survivors in "a 15-foot dome with 6,000 LED lights and 116 cameras, and asked them 1,000 questions about their life over five days." As Smith noted, "Just seeing that, that would be an imposition and possibly dangerous for that individual to go through that experience. . . . We ensured that first of all in the very beginning of this project, we engaged with Holocaust survivors on our advisory group, before we even went into pilot. . . . In fact, before we went into prototyping, survivors went down to look at the studio, talk through the process of production, [and] think through the issues that might come out of it."

Scholars and activists such as Sasha Costanza-Chock have expressed concern for the abandonment of some ethical and equitable concerns in emergent tech spaces, issues that have been raised and at least partially addressed in conventional documentary and journalism.

Emerging tech is accelerating and fragmented. It affords new opportunities for expression and communication but also introduces dilemmas of limited distribution and, in many cases, technology-first agendas. Along with the struggle to arrive at coherent approaches to these immersive forms, makers concerned with community-based co-creation must confront the inaccessibility of the devices and knowledge they are founded on.

SPOTLIGHT: *QUIPU PROJECT*—PERUVIAN STERILIZATION SURVIVORS SPEAK

by SARA RAFSKY

The stories are painful to hear. In recording after recording, the subjects speak about being among the nearly 300,000 women (along with thousands of men) who were brutally subjected to sterilization under the government of former Peruvian president Alberto Fujimori in the 1990s. The program targeted rural, poor, and Indigenous communities, who were forced to participate without their informed consent. Many continue to suffer painful symptoms and side effects from the procedures to this day. Their testimonies, along with listeners' responses, form the core of the interactive and co-creative online documentary *Quipu Project* (2015).

Filmmakers Rosemarie Lerner (Peru) and María Ignacia Court (Chile) were galvanized into action when the sterilization program was debated during the 2011 Peruvian presidential campaign. Lerner realized that while many of the survivors had

Participants in Peru engage with the Quipu Project *(2015) on their mobile phones. Used with permission from María Ignacia Court and Rosemarie Lerner.*

been interviewed by other media, they had never had an opportunity to participate in or comment on the final productions. According to a study by Mandy Rose, the two filmmakers set out to develop a co-creative approach as a means "to engage with this disenfranchised community in a way that would offer them control in the process and would prove of value to them in their struggle for justice."

In the ancient Incan system of Quipu, colorful knotted strings were used to keep official records and tell stories. In the *Quipu Project*, users click through colored-dot icons, each representing a section of the testimonies of the more than one hundred women (and counting) who dialed a free phone number and left anonymous messages about their experiences. People from around the world who listened to their stories then recorded messages in response. The audio was collected using Drupal VoIP, an open-source, voice-over-internet protocol (VoIP) developed at MIT's Center for Civic Media. The result is a participatory oral history project that culminates in a call to action.[27]

Lerner and Ignacia Court spent over a year traveling to various regions of Peru, gradually and with setbacks gaining the confidence of local women and civil society organizations and recording stories. The co-creators had to take into account the limited online connectivity of the region. "The only technology [the participants] had was radio and telephone. So we realized that we had to use what they used if we really wanted to tell their stories," Maria Ignacio Court said in a 2016 interview.[28] In the end, Lerner and Court devised a system that blended low-tech phone technology for recording and a high-tech digital interface for the user experience. They equipped some women as story hunters to seek the widest possible group of participants. Later, VoIP technology would allow participants of the project to listen to all the messages and be notified when a listener had responded. It was hoped that this process would help build a broader mutual support network for these survivors. Hearing others' stories offered the women a way to "break through the isolation and stigmatization they had faced in the past," according to the team in an i-Docs interview.[29] The anonymity of the phone-recording process was crucial to protect the survivors in a still-dangerous context.

Participants gather in the Quipu Project *(2015) in Peru. The co-creators hoped that the process would help build a broader mutual support network for these survivors of forced sterilization. Used with permission from María Ignacia Court and Rosemarie Lerner.*

With the executive producer Sebastián Melo and creative technologist Ewan Cass-Kavanagh, the *Quipu* documentarists sorted testimonies about the sterilization program into different classifications. These were stories about the actual operations, life afterward, and survivors' and activists' search for justice. The web documentary, which was launched as a pilot project at the 2014 i-Docs conference, was exhibited at the 2015 IDFA Doclab and won a legacy Peabody award in 2022 for it's contribution to the field. With funding raised via crowdsourcing, the project later expanded to include a linear twenty-minute documentary on *The Guardian* website in 2016.

"We didn't want to tell a story about those affected," Lerner told Rose. "We wanted to tell a story with them . . . where people could share their own stories in their own voices, but they could also experience the interactive documentary and become its first audience." For the survivors, sharing their stories functioned as a type of rehearsal for potential court testimony, another manner in which co-creative media making could serve as an agent "in the fight for justice."[30]

SPOTLIGHT: *QUESTION BRIDGE*

by SARA RAFSKY

"How do you know when you become a man?" is a daunting question asked by eight-year-old Josiah Yoba but one that the adult Gavin Armour attempts to tackle. "You don't wake up one morning and become a man," Armour answers. "It's your experiences and how you react to them." The exchange is poignant, but neither Yoba nor Armour are talking directly to each other; rather, they are speaking to cameras. The way their filmed comments interact is at the crux of *Question Bridge*, an ambitious, multiyear transmedia project about Black male identity. *Question Bridge*, according to its mission statement, "facilitates a dialogue between a critical mass of Black men from diverse and contending backgrounds and creates a platform for them to represent and redefine Black male identity in America."

This is how the project developed: Black men were recorded asking questions they would like to pose to other Black men. The footage was then played to other Black men who were filmed responding and asking their own questions, and the chain continued. The participants were diverse—young, old, professionals, prisoners, celebrities—and the questions ranged from sweet ("How do you know she's the one?") to political (discussions around the use of the N-word) to existential ("I wonder, Black man, are you really ready for freedom?"). The result is a co-created mosaic portrait that defies any one narrative of Black male identity.

An installation view of Question Bridge. *Used with permission from Kamal Sinclair.*

Question Bridge originated as a project by artist Chris Johnson, who in 1996 used the same technique to record the multiclass and multigenerational African American community in San Diego. The piece was reborn over a decade later when artist Hank Willis Thomas suggested to Johnson that they revisit the project, this time focusing specifically on Black males. Johnson and Thomas joined forces with artists Kamal Sinclair and Bayeté Ross Smith and traveled to nine cities across America, where they recorded over 160 men asking and responding to more than 1,600 questions. As Sinclair told us, "There is no monolithic Black male identity. By inviting people that identify as Black and male to participate in a dialogue and to define themselves in their own terms, you break open the constraints on Black male identity and expose an incredible diversity of thought. People feel liberated from these limiting ideas of what it means to be Black and male."

Encouraging the participants to craft their own questions enabled participants' agency over the process and resulted in them breaking from traditional interview frameworks. The indirect nature of the filmed and assembled dialogue was critical, the creators wrote on the project's website, as it "reduces the stress of normal

An installation view of Question Bridge. *Used with permission from Kamal Sinclair.*

face-to-face conversations and makes people feel more comfortable with expressing their deeply held feelings on topics that divide, unite and puzzle." As a second layer, the creators hoped that by exposing viewers to the multifaceted thoughts and opinions of Black men, stereotypes would be deconstructed, and the project would enable viewers to overcome negative bias toward, "arguably, the most opaque and feared demographic in America": Black males.

The five-channel video installation launched at the 2012 Sundance Film Festival and went on to tour more than sixty festivals, museums, and institutions. After a 2013 Kickstarter campaign that raised US$75,000, the project expanded to include a website and mobile app for users that enabled the addition of more content. The program also developed into a high school curriculum and an Aperture Foundation photography book,[31] according to a case study published by Media Impact Funders.[32] The organizers hosted more than a dozen Blueprint Roundtables, multigenerational community engagement events that seek to identify "road maps to success" for Black men and boys. In 2015, *Question Bridge* was the recipient of the International Center for Photography's Infinity Award in the new media category. A year later, a video,

A live panel of Question Bridge. *Used with permission from Kamal Sinclair.*

three and a half hours long, from the project was added to the permanent collection of the new Smithsonian National Museum of African American History and Culture.

CO-CREATION IN JOURNALISM

Not too long ago the term *collaboration* was something of a dirty word in newsrooms. Immersed in an economic and cultural model that for a century had rewarded scoops and exclusives, journalists and editors thrived on competition, while established ethical standards kept them removed from their subjects. Yet in recent years as changes in technology and the marketplace have upended the news industry, the collaboration prohibition has become one of the many edicts in journalism to be challenged. Media outlets have experimented with new journalistic forms ranging from big data dives and visualizations to games and interactive documentaries. Many of these practices as well as more traditional reporting have been supported by co-creative methods. Of course, traditions of attribution, reputation, credibility, professionalism, and credentials remain important in journalism, an area that demonstrates the advantages of authorship. But perhaps there is no need for a binary opposition between these values. Perhaps there is a third way.

"The essence of the current infrastructure for journalism is that it's built on col-laboration, even if that collaboration is implicit," said Carlos Martinez de la Serna, the program director of the Committee to Protect Journalists. "We might not even be realizing the full extent of the role of collaboration, but it's there. It's about the networked world we live in."

Martinez de la Serna argues that human-centered design snuck in the back door of journalism through the ways that infrastructures and interfaces are built. For over a century in print, radio, and TV, design was mostly thought of as graphic design, but today's journalism is based on digital platforms, which are in turn best created with human-centered design. "We are seeing people building interfaces for news and testing those interfaces with users and incorporating their feedback, and they cannot think of other ways of working. So we are seeing co-creation happen in journalism. Again, it's not explicit. Maybe we're not even realizing it's happening. But it's happening."

The role of community in journalism is not new. Martinez de la Serna points to important conversations on the subject with the rise of the "public school of journal-ism" in the 1980s. Jessica Clark, director of the Philadelphia-based Dot Connector Stu-dio, recalled those efforts as "the precursor to all this online stuff. Failed. It was an idea advanced by very well-meaning academics and funders over the course of a decade or so, and of course the big newspapers never really picked it up, because they didn't have to, right?"

Today, many media organizations are embracing what Andrew DeVigal, the chair of journalism innovation and civic engagement the University of Oregon School of Journalism, has dubbed a "continuum of engagement."[33] This concept moves beyond the social media outreach of most media outlets in order to get their public to see and share content or "transactional" engagement, as defined by DeVigal. Instead, they embrace a more "relational" approach. Projects range from opening up the process of receiving story ideas to experiments in participatory reporting and even joint author-ship with subjects.

Clark suggested that in this environment, a new brand of "engaged journalism" is emerging. She has been involved in aggregating engagement case studies for DeVi-gal and the University of Oregon Agora Journalism Center's platform called Gather, funded by the Knight Foundation, Democracy Fund, and other backers. As in the documentary field, engaged journalism practices range from the participatory to the interventionist.

At the *New York Times*, Malachy Browne (formerly of Storyful) has been crowd-sourcing to push the bounds of what is possible with co-created, investigative journal-ism in a team called Visual Forensics. For instance, Browne draws on vast sources of eyewitness video, tweets, and social media data (along with their metadata) to piece together complex events such as climate disasters and mass shootings, as in his award-winning *10 Minutes. 12 Gunfire Bursts. 30 Videos: Mapping the Las Vegas Massacre* (2018). Browne calls this methodology "forensic journalism." When the team finds a shot from a civilian phone camera panning across a wall, they will enhance the clock on the wall to secure a time reference. They can sync thirty videos by using the sounds of the gunshot blasts to place the data, collected from multiple sources, in a coherent timeline.

Browne furthers his co-creative approach as he develops relationships with phone- and camera-wielding civilians on the ground in quickly evolving situations; this helps him coordinate livestreams, the collection of evidence, and interviews. In Houston during the flooding after Hurricane Harvey (2017), for example, he teamed up with Mattress Mack, a local who opened up his warehouse and lined the floor with mattresses to welcome people in need of shelter. Mack helped facilitate live FaceTime interviews with his guests, with direction from the team from the *New York Times*. This echoes Sam Gregory's Witness organization and Michael Premo and Rachel Falcone's *Sandy Storyline* but from across the journalistic divide.

Before he joined the *New York Times*, Browne undertook a complex investigation using his own social networks, including Human Rights Watch, when he received a tip about bombs built in Italy being smuggled through Saudi Arabia to be dropped on Yemen. Together, a networked community traced the routes the bombs were taken on, photographing ships and planes, documenting shipping logs, and communicating live over Twitter. Once the story went viral and politicians began discussing the issue in the Italian parliament, another part of the network would translate live from Italian to English. As Browne explained, "The community emerges, and we all work together and bring our own strengths to it and actually get inside something like that. . . . That was very much co-creative, collaborative, very transparent and very much in the public sphere. We would be chatting privately, but we would be posting what we're seeing onto Twitter and sharing it to different threads."

"The real value and the strength of social networks is in hard-to-reach places, or places where there is breakdown of the media, independent media," said Browne. He gave examples of remarkable social media and blogging efforts such as SOS Media Burundi and Raqqa Is Being Silently Slaughtered, both of which leveraged the live and decentralized nature of the web to keep the eyes of the world watching and holding aggressors accountable.

Local co-creative journalism is not only important at the front lines of war and violence. Arguably, local journalism everywhere has been co-creative and community-based since the first edition of the first newspaper. Sam Ford, executive director at AccelerateKY, looks back to his beginnings in community-based journalism while growing up in Ohio County, Kentucky: "There's been a long-standing tradition of deep audience participation in creating the news. I got my start in journalism at 12, writing a society column that runs on the women's features page in these rural newspapers. Every rural community of 150, 200, 300, 400, 500 people had their own. *McHenry News* was the name of mine. There was the *Rosine Happenings*. There was the *No Creek News*. Every town had their own column. You had a curator . . . who gathered the information, but those columns were a communal space."

The same has often held true of African American community newspapers and many ethnic or linguistically specific publications.

Even today, Ford takes issue with the claim that a dearth of reliable journalism outside of major urban areas makes "news deserts" of large regions of the United States. "There are a lot of news organizations already at work at the hyperlocal level in these places," he said. These include the Solutions Journalism Network, whose methods may be based on conventional reporting but whose fundamental approach is based on the co-creative principle that answers are found within communities. The network trains journalists to cover how people on the ground are responding to problems and collaborates with hundreds of news organizations to pursue solutions-oriented projects.

Al Jazeera's *The Stream*, alternately, is a broadcast and internet program that uses information extracted from online hashtag activist movements; disseminates callouts and augments unfolding news through social media, talk show formats, and interviews with experts; and includes news segments produced by citizens and activists.

Journalists are also finding ways of working with communities beyond the digital divide. Julia Kumari Drapkin, the founder of ISeeChange, makes efforts to co-create with the most vulnerable people affected by climate change, although this means crossing the lines of conventional journalism. "Reaching people who aren't necessarily online, how do you do that? Working with community partners. That is one of the first taboos we have broken for traditional journalism. . . . Partnering with the community suddenly lands you in a place where you're an advocate. . . . People look at us and they say, 'Oh, they're advocates,' and we're like 'Actually, we're just 21st-century journalism, and this is 21st-century science.'"

For a project on urban heat, ISeeChange partnered with community groups to place heat sensors in thirty public housing households in Harlem. The data generated, along with the stories told by community members, helped to create a much more reliable picture of the trapped urban heat that residents endure in the summer

FIRST SNOW REPORTS 2018

❄ Earlier than average   ❄ Later than average   ❄ Average

NOVEMBER 9
Franklin,
Pennsylvania

OCTOBER 16
Wheat Ridge,
Colorado

NOVEMBER 9
Kansas City,
Missouri

OCTOBER 20
Middleton,
Wisconsin

OCTOBER 6
Mt. Lamborn,
Colorado

NOVEMBER 5
Highlands Ranch,
Colorado

NOVEMBER 15
Ambler,
Pennsylvania

NOVEMBER 13
Fairbanks, Alaska

Averages source: The Weather Channel

A web screenshot of ISeeChange, a platform and global community that uses co-creative and cross-disciplinary methods between journalists, scientists, on-the-ground community members, and now with NASA to monitor and understand the socioenvironmental impacts of weather and climate change. Used with permission from Julia Kumari Drapkin.

in Harlem. The city credits the project for changes in policies around cooling centers and public heat advisories.

"We need a different framework," Martinez de la Serna told us. "The commons give us a different way of understanding . . . journalistic relations. The commons are, fundamentally, based on co-creation. There's no commons if we are not equals, even if we have different roles. But we own this space. We co-own this space."

The commons is key to co-creative journalism and can help relieve the tension between authorship and co-creation. Nothing is cut and dried, yet in the digital age there is an increasingly evident direction and a redefinition of the journalistic commons, connecting it to other forms of commons and community-based methods, with governance as a key part.

SPOTLIGHT: COLLABORATIVE JOURNALISM

by SARA RAFSKY

BETWEEN NEWSROOMS

The most wide-reaching, well-developed, and common form of collaborative journalism to date is collaboration between newsrooms. When the *New York Times* and *ProPublica* started collaborating on investigations and sharing bylines nearly a decade ago, it signaled a major cultural shift for the most prominent US news outlets. Now these kinds of investigations are myriad and include some of the most celebrated reporting of recent years such as the famous Panama Papers and Paradise Papers exposés of tax havens.

Recognizing its significance, the Center for Cooperative Media at Montclair State University has been compiling a comprehensive database of collaborative journalism projects from around the world.[34] The center's research has identified six distinct models of collaborative journalism that "span collaborations from hyperlocal to the international levels" and are based on how long organizations worked together and how they integrated their work and their processes.[35]

This model of collaboration has included partnering between newsrooms seeking to cover broad, overarching issues as with the Electionland project, which covers voter access and suppression issues, as well as examples of national news outlets seeking to amplify the work of local reporting, such as the *New York Times* collaboration with various local news organizations to expose racial divides in regard to COVID-19 rates and nursing homes.[36] As the practice has spread more widely, new infrastructure is being developed to facilitate it. The open-source Project Facet, for example, helps newsrooms manage planning, organizing, storage, and communication needs across different platforms.[37]

PROJECT-BASED MEDIA PARTNERSHIPS

Legacy media institutions have also reached across the divide between media forms in co-creative ways in the last decade to pool resources, reach wider audiences, and invent new strategic transmedia spaces. This cross-collaboration is evidenced in the MIT Open Documentary Lab's report *Mapping the Intersection of Two Cultures: Interactive Documentary and Journalism.* "The current transition, for all of its disruptions, offers ways to make fuller use of journalistic archives, audiences as partners, and new and immersive story techniques. Embracing change is rarely easy, but the stakes for

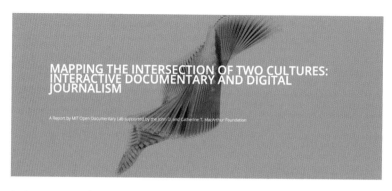

Mapping the Intersection of Two Cultures: Interactive Documentary and Digital Journalism. *Used with permission from the MIT Open Documentary Lab.*

informed civic participation are too important for business-as-usual, and the potential rewards are too ripe to ignore."[38]

Detailed case studies in the report outline a variety of collaborations between newspapers, digital publications, broadcast television, and film producers. An early example came during Cynthia Lopez's time as commissioner at PBS's *POV*, where she led a partnership between public television, Oprah Winfrey's Harpo Studios, and ABC's Ted Koppel for *Two Towns of Jasper* (2002). The film was made through two perspectives, one White and one Black, to document the 1998 lynching of James Byrd in East Texas and resulted in a week of *Race in America* programming that reached an estimated sixty million people across the country, according to Lopez. "[We] brought together community groups, journalists, independent filmmakers, and educators to be part of the conversation."

Since then, short-term project-based co-creative experiments across legacy institutions have become not only feasible but even routine. *Post Mortem*, a yearlong exploration of death investigations in America, for example, aired as a multipart radio series on NPR, an hour-long *Frontline* documentary on PBS, and a detailed report on ProPublica.org. And in 2021, for example, PBS, under the guidance of Opeyemi Olukemi, then digital vice president, collaborated with Instagram and the NFB on experimental projects called *Otherly*.

WITH OTHER DISCIPLINES

Collaborative journalism can also include partnerships between journalists and experts working in other fields. While the media have always sought out these groups as sources for their reporting, under this model reporters and experts from other disciplines work together to define reporting priorities. These methodologies and research protocols are less developed, but much of this collaboration has happened at the academic level.

Between 2014 and 2016, for example, the *Chicago Tribune* worked with data scientists, pharmacologists, and cellular researchers at the Columbia University Medical Center to identify prescription drug combinations that can lead to potentially fatal heart conditions. The results were published in news stories by the *Tribune* and in academic papers by the scientists.[39] In an article about the process, *Tribune* reporter Sam Roe recommended that journalists seeking to experiment with similar efforts should choose partners with a "similar commitment to objectivity and truth," set standards for successful outcomes, and maintain open and ongoing dialogue with all participants.[40]

At the Global Reporting Centre at the University of British Columbia, academic researchers have been brought into the newsroom as ad hoc editors, helping journalists report more accurately on the researchers' areas of expertise.[41] The website *Reveal*, which produces podcast and provides a social media platform for the Center for Investigative Reporting, organized the Mind to Mind Symposium with Stanford University to bring journalists and academics together to discuss new ways to partner and "fill the gap or to look at under-researched areas."[42]

Outside of academia, there is growing collaboration between media outlets and the tech sector. In addition to instructing and informing journalists about these areas, people in tech and computer science have worked with journalists to create tools to facilitate the reporting process, often in the form of hackathons for professionals and specially designed student courses.

BETWEEN JOURNALISTS AND SUBJECTS

Finally, there is collaboration between journalists and the communities they cover, sometimes known as "engagement journalism." This is perhaps the thorniest and most complex area of collaboration, as it must contend with obstacles ranging from matters of editorial ownership and control to deeply held conventions and ethical principles about "objectivity" and the necessary separation between journalists, their sources, and their audiences.

Nevertheless, examples of participatory reporting at the national and international levels include several high-profile projects from *The Guardian*. For "The Counted," for example, *Guardian* journalists combined their own reporting with verified crowd-sourced information about people killed in the United States by the police so as to create a more comprehensive database.

Further along the "continuum of engagement" identified by DeVigal are media outlets that seek to integrate the expertise and priorities of the communities they serve more directly into story generating and reporting, putting reporters and their subjects on a more equal footing.

Projects such as *Sandy Storyline*, *18 Days in Egypt*, and *99%: The Occupy Wall Street Collaborative Film* have explored the form of crowdsourced documentaries. In addition to creating an interactive web documentary about a rural community in West Virginia, the creators of *Hollow* helped set up a community newspaper/blog with residents. The Seattle Times Education Lab has held community brainstorming sessions with parents, students, teachers, and education advocates and experimented with new ways to feature community voices, including live chats, reader questionnaires, and regular guest columns.[43] And Chicago-based public radio station WBEZ created Curious City, a crowdsourced news platform where listeners suggested and voted on story ideas and sometimes contributed to the reporting itself. Curious City was part of Localore, a co-creative initiative run by AIR (a global community of independent audio producers) across a cohort of public stations to spark community-first, cross-platform storytelling.

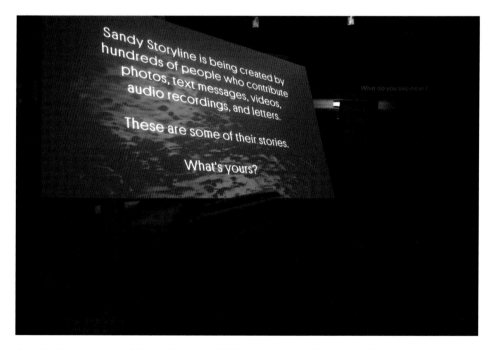

A projection screening of Sandy Storyline *(2012), a co-created documentary by Michael Premo and Rachel Falcone. Used with permission from Micheal Premo.*

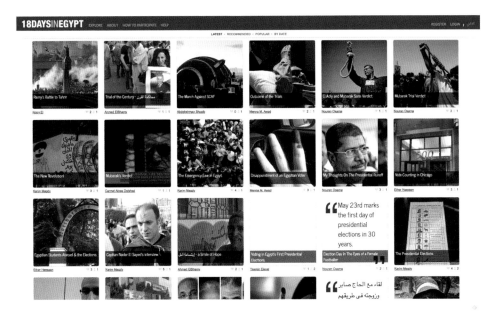

A screenshot of 18 Days in Egypt *(2011). Photo courtesy of Jigar Metha and Yasmin Elayat. Used with permission from Yasmin Elayat.*

As with collaborations across newsrooms, new infrastructure is emerging to facilitate community engagement journalism. The Chicago-based company Hearken builds on the success of Curious City and offers a customized platform and editorial framework that enables journalists to better partner with the public for each step of the reporting process.[44] The Freedom of the Press Foundation's Secure Drop project provides a means for whistleblowers to leak documents to the press anonymously and to increase accountability while reducing fear of reprisals.[45] And in October of 2017, DeVigal and his team at the Agora Journalism Project at the University of Oregon launched the aforementioned Gather, a platform that brings together those working in the field and shares resources, case studies, and best practices.[46]

DESIGNING FUTURES

In the history of documentary and community-based media making, co-creation practices are those that center the people most affected. They could be the users of a tool, the readers of a nonlinear interactive online story, the participants in a story

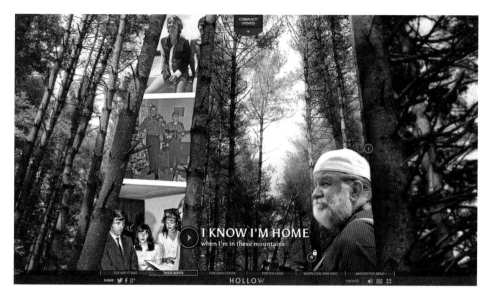

A screenshot from the web documentary Hollow *(2013) combined in-person co-creation methods with online ones (2012). Used with permission from Elaine McMillion Sheldon.*

world—or metaverse, or people living on the front lines of complex problems. Practical approaches vary substantially, from those of nonhierarchical art collectives to more interventionist strategies to co-designing change.

None of this work within communities, however, exists outside the reach of technologies and their encoded inequities—from early cameras to the internet to VR headsets and increasingly algorithmic systems. Race itself is a technology, as Ruha Benjamin points out.[47] Matrices of relations and systems of power run throughout the genealogy of documentary, journalism, and nonfiction.

How do we unshackle media making from these oppressive systems and move toward, as the Detroit Narrative Agency says, co-liberation?

Community-based values and practices live at the core of co-creation in documentary and journalism. But they function beyond the frame of a naive perception of media making that still pervades the industry: the simplistic triad of media producers/subjects/audiences. The aperture needs to open far and wide to locate these practices within the webs and perils of the platforms and tech systems that increasingly permeate all domains of our lives. As Duke Redbird says, these systems, all framed as consumerism, "mine our interior arenas."[48] COVID-19 in particular has pried open the cracks as people have struggled to transpose complex physical communities to entirely online experiences.

As media makers, we need to not only tell and expose these big stories but also co-create with many movements to confront and, urgently, build alternatives to these toxic and harmful systems.

Tim Berners-Lee, the inventor of the World Wide Web, has called for a redecentralization of the internet. He and coders around the world are co-creating a new set of software, Solid, in an attempt to reclaim the internet from the new tech behemoths. On his World Wide Web Foundation's site, Berners-Lee wrote an open letter in 2018, saying "while the problems facing the web are complex and large, I think we should see them as bugs: problems with existing code and software systems that have been created by people—and can be fixed by people."[49]

Ruha Benjamin, however, implores us to see that many of these so-called bugs, glitches, and fixes are precisely part of the system—they are features. So, what next? Deibert argues in his 2020 book that we reset: turn the system off and start up again, slower, with regulation, reform, retreat and restraint. Benjamin draws on the framework of abolitionism to combat coded inequity. She points to digital justice movements that audit AI, democratize data, and demystify tech and she that says the answers are not merely computational. "*Narrative tools* are essential," she insists.[50] Sasha Costanza-Chock similarly advocates for design justice and shows how it can lead to new structures.

WITNESS continues to model trail-blazing community-centered work in its deep support of human rights movements and its interrogation of platforms, deepfakes, and accountability in oversight at platforms and policies at government levels. Thenmozhi Soundararajan and Equality Labs embed digital security directly within artistic and political work. And Duke Redbird suggests that the way forward is found within Indigenous epistemologies: "Let us not forget that the origin of all knowledge and wisdom comes from a platform called Mother Earth."[51] He says of the current moment that "the upside is that we have an opportunity now to share from the Indigenous communities around the world a long tradition of living in communion with nature and bringing that sensitivity and sustainability to a modern world."

These are all crucial parts of the tool kit we need to assemble collectively so as to build a just future. The climate movement has taken fifty years to reach a stage of mass global youth protest movements and widely accepted Green New Deals. We don't have fifty years to confront surveillance capitalism and the New Jim Code. "Nothing about us without us" takes on new meaning in this matrix. This story is yet to be written, but we can learn from the vast historical and contemporary canon of co-creative media making grounded within both real-world communities and those online. We can also learn better how to co-create across disciplines and with systems, nuance our understanding of them, and change them.

3

ESTUARIES: CO-CREATING ACROSS DISCIPLINES AND ORGANIZATIONS

It's the classic picture of a Hollywood set: The director yells "cut." The producer yells "fired." In conventional hierarchical media-making teams, a rigid chain of command dictates who makes decisions and when and how so as to make the process as efficient and linear as possible. Every role has a job description and very specific expertise. Similarly, in engineering, technology, and science, the modus operandi is to isolate a specific problem, and develop a specific solution using a streamlined hierarchy of roles and efficient methods within a specific discipline.

But the stories we tell, the technologies we use, and the problems we face and why they matter are all incredibly complex, bound up in large systems and frameworks. No one person, discipline, or organization has the capacity or knowledge to tackle it on their own. We need new ways of working across systems, and media co-creators of all kinds have been charting the course.

In co-creation, when teams cross disciplinary lines, they embark on parallel paths of discovery rather than ranking one discipline's priorities over the other. It's similar for organizations. In working together, they agree to write new rules on how a project or relationship might emerge.

Professional disciplines, sectors, and organizations are often deeply divided by specialization, jargon, protocols, and hierarchies. But around the world, artists, journalists, scientists, and institutions are developing forms in which to mix and create ideas and worldviews through crossing and dissolving these boundaries, comparable to the

Collage created by Helios Design Lab. Photo courtesy of Marisa Jahn.

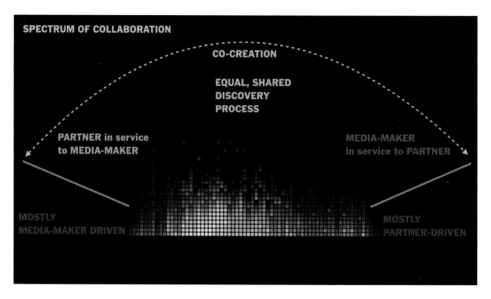

Spectrum of collaboration. Graphic used with permission from the MIT Co-Creation Studio.

way that freshwater and saltwater mix in natural estuaries. With unconventional work methods, the sectors have arrived at unexpected outcomes. From temporary hackathons to more sustained models such as residencies, incubators, platforms, and labs, they might mix art, science, journalism, and human-centered design to produce, for example, climate change reporting informed by citizens' collection of evidence. These experiments tackle the complex problems of the century while enabling inclusivity and diversity and honoring the expertise of people from all walks of life.

By our definition, media co-creation lives within a specific range. It is work that is devised together, embarks on a joint journey of discovery, and has outcomes with implications for multiple disciplines and areas of knowledge. Co-creation can work in cross-, trans-, and even antidisciplinary spaces. But it takes specific conditions to flourish: openness, flexibility, intention, iteration, abandonment of ego, deep teamwork, and, most significantly, time.

BUILDING AND TEARING DOWN SILOS

Modern academic disciplines and their attendant professional societies emerged with the secularization of universities in the nineteenth century. The process largely coincides with the "invention" of national traditions (e.g., folk costumes,) and even the pope's "infallibility." It's a curious epoch of reification. Before then, the European university of the late Middle Ages through the early nineteenth century was organized along sectors such as canon law, theology, the arts, and medicine and before that the trivium (grammar, logic, and rhetoric) and the quadrivium (arithmetic, astronomy, geometry, and music). Despite these undulations and despite their recent vintage, our current academic silos often feel as though they are carved in stone.

This left a mark on how the arts and sciences are imagined. When specialists enter cross-disciplinary settings, there are legitimate fears of a backlash in the form of the braking mechanisms of hiring and tenure decisions at universities, grant categories and review boards, and academic journals and editorial boards. Still, collaboration across disciplines never has been abandoned completely. We see efforts such as multidisciplinary teams in medicine, multiorganization collaborations in other fields, and attempts to create interdisciplinary spaces.

On April 10, 1861, the governor of the Commonwealth of Massachusetts signed a charter for the incorporation of the "Massachusetts Institute of Technology and

Building 20 at MIT was hastily constructed in 1943 as a temporary structure in the emergency war effort to house a new secret lab called the Rad Lab. The building lasted until 1998. Many researchers credit its ad hoc nature for major discoveries and innovations in multiple fields. It was a place for high-risk projects, collaborations, and experiments that benefited from unofficial DIY removal of walls and the running of wires down hallways along with physical and intellectual improvisation across disciplines. Image made available by the MIT Museum under Creative Commons.

Boston Society of Natural History." MIT broke down barriers between areas of classic scientific knowledge as its founders sought new educational models that stretched beyond the confines of institutions that they claimed no longer served an industrializing society. The motto of MIT is *mens et manus*, "mind and hand" (i.e., learning by doing). While it was a long while before the institute began work that intersected with the arts and humanities—much of its agenda was driven for a time by military funding—MIT's example shows how thinking can be affected by diverse disciplinarity. By the 1930s, the fruits of that strategy appeared in such developments as the Radiation Laboratory and cybernetics. MIT understood that to spark innovation, diverse teams of scientists needed to work in the same physical spaces and easily share ideas, information, and projects.

As the twentieth century progressed, so too did awareness of the advantages of collaboration across disciplines and borders. Yet, the scientific world was still missing a tool for widespread instantaneous collaboration across geographical space, one that would be able to accelerate the now well-known benefits of multidisciplinary approaches in the sciences. The internet (known then as ARPANET) had been invented in 1969 but lacked a many-to-many platform. The development of the web as we know it today was a luminary example of co-creation: the invention of a system that would allow multiple players moving in uncharted space to develop, iterate, and build platforms for sustained collaboration.

Tim Berners-Lee, now a professor at MIT, was a young engineer freelancing for the CERN laboratory in the latter part of the twentieth century. CERN had been created jointly by a large international consortium of research centers, but communication was constrained by existing one-to-one and one-to-many media systems. Berners-Lee developed what would eventually be known as the World Wide Web, a peer-to-peer network of nodes built on top of the internet. Fortunately, he also felt that it should be free and accessible to the rest of the world, and with the generous act of releasing his code in 1991, he gave the world a decentralized computer networking protocol to share, innovate, and eventually start up at levels not seen before.

Unfortunately, most cultural institutions around the world were slow to dive into the digital era that the open web revealed and only did so by creating siloed digital departments. They tended to relegate use of the web to marketing and only later hired a few digital producers. "Sometimes there's a department 'over there,' a skunkworks, a secret department, a leading-edge technology department, and they're responsible for innovation," said Gerry Flahive, a documentary producer and writer who worked at the National Film Board (NFB) of Canada for over three decades. "Meanwhile, the rest of us are over digging coal just the way we have for a hundred years." In the past decade, legacy institutions have finally come around to understanding that digital infrastructures and cultures are not only native platforms on which to create but have also transformed every possible form of cultural production profoundly.

Meanwhile, Silicon Valley and the larger tech industry have taken advantage of that lag to become the twenty-first century's predominant drivers of innovation. But they too suffer from hyperspecialization; they tend to bring little social or cultural context to their work and neglect interdisciplinary practices beyond the fields of engineering, programming, marketing, and complex finance. The growing body of critiques that expose the harm and oppression embedded in their data, platforms, and industry has been met with astoundingly naive or disingenuous surprise among these corporations' senior management.

"I'd argue that if Facebook had worked very deeply and collaboratively with the creative community ten years ago or so, they possibly wouldn't be in the trouble that they're in currently," said Domhnaill Hernon, an engineer who ran the Experiments in Art and Technology (E.A.T.) program at Nokia Bell Labs. "Nor would any of the other big tech companies, in how they develop their technologies. So that's one element or one dimension—how they [could] influence our thinking to be more human-centric, which is completely, absolutely missing from the entire STEM education system globally as best I can tell. And that has not changed significantly in quite some time."

Others argue that it's not an omission; instead, it's that the platforms were designed that way. Evgeny Morozov describes this gap as "techno-solutionism," that is, the troublesome belief that all problems can be solved by technology. "[Solutionists] have a very poor grasp not just of human nature but also of the complex practices that this nature begets and thrives on. It's as if the solutionists have never lived a life of their own but learned everything they know from books—and those books weren't novels but manuals for refrigerators, vacuum cleaners, and washing machines."[1]

At the moment, Silicon Valley's latest round of solutionism includes the creation of oversight committees, ethical artificial intelligence (AI) teams, and other substructures that seem superficial and performative. Many experts believe that effective self-regulation within Silicon Valley is just not possible. In a 2021 exposé of Facebook's "Responsible AI" team, Karen Hao illustrates how any meaningful oversight and ethics of AI systems fly in the face of the company's core mission, "[Mark] Zuckerberg's relentless desire for growth."[2]

The tech industry is building the new engines and operating systems of humanity, but when they are exposed they are riddled with systemic oppression. Ruha Benjamin

Making a New Reality. *Used with permission from Kamal Sinclair.*

exposes the New Jim Code in *Race after Technology* (2019). She argues that racism has not disappeared but instead has evolved, as it is built, designed, and coded directly into our technological systems to reproduce and reinforce segregation and racial violence in the twenty-first century. Virginia Eubanks in her book *Automating Inequality (2018)* describes how high-tech tools profile, police, and punish the poor. While companies such as YouTube, Spotify, Amazon, and Google seemingly provide benign new platforms for the arts and culture, Silicon Valley primarily imagines culture as an engineering problem that can be addressed with simple recommendation algorithms that stand in for individual and group tastes. Benjamin calls for an abolitionist approach to these systems and insists that narrative tools are instrumental in the fight.

In *Making a New Reality* (2020), a report on inclusion and diversity in emerging tech storytelling, Kamal Sinclair, executive director of the Guild of Future Architects, calls for sharing space. "Promote the intrinsic value of each other's sectors. Create shared language and practices. Invest in hybrid talent. Strategically embed artists with technologists and scientists. Thoughtfully include the arts in spaces of power. Build an interdisciplinary community outside spaces of power."[3]

CO-CREATION AMONG ARTISTS, SCIENTISTS, AND MEDIA

"Science is looking for answers, while art is looking for questions," artist Marc Quinn told Stuart Jeffries of *The Guardian* in 2011 while on a commission from England's National Portrait Gallery to make a portrait of John Sulston, one of the scientists who decoded the human genome.[4] Quinn used DNA from samples of Sulston's sperm and then cut and grew them in bacteria. But increasingly, artists and scientists have been discovering that richer outcomes can result when neither partner has all the answers or even all the questions.

Gina Czarnecki, a British pioneer in bioart, considered most of her earlier partnerships with scientists as collaborative rather than co-creative, treating the researchers mostly as "people who provide information." But for her 2016 project *Heirloom,* she came together with University of Liverpool professor of clinical sciences John Hunt in a deep, co-creative process of discovery. Using cells harvested from Czarnecki's daughters, they built 3D masks that have traveled to major galleries, but the process also resulted in Hunt inventing a new medium for growing stem cells. In previous clinical applications, cells were grown flat and were then difficult to stretch over curved human surfaces. Face Lab at Liverpool John Moores University is setting out to explore how the new techniques might be applied to reconstructing faces from archaeological finds, and there is growing interest in using these methods to treat burned and scarred patients.

Installation view of Heirloom *(2016), Gina Czarnecki and John Hunt. Used with permission from Gina Czarnecki.*

"Working with John," Czarnecki told the *Forma Arts* interviewer, "has very much been an equal co-authored collaboration, with the focus being on progressing not only the artistic concerns which helped scientists with public engagement and educational agendas, but also the scientific possibilities which—working in the context of art—could facilitate new hybrid discoveries at a different pace."[5]

Lisa Parks, a digital media scholar, first experienced co-creation in 2003 when she worked with two artist-academics to produce research on physical and media infrastructure in Europe and for her critically acclaimed art shows in Berlin. "We were excited about issues of technological literacy," she said, "and how people in relation to these systems understood them as part of their life worlds, and what the limits of their knowledge [were]." In Parks's latest project, she is studying local responses to the centralization of digital infrastructures through what she terms "Network Sovereignty." She is co-creating with Ramesh Srinivasan at UCLA along with three community partners around the world: Rhizomatica, a publicly owned mobile phone network in the mountains in Oaxaca, Mexico; a community internet network in the Serengeti, Tanzania; and Oki Communications, which is partly owned by the Blackfeet Nation. As she observed in our interview, "There's reason for those on the fringes and in these rural areas to actually assert ownership over these systems rather than to just

A screenshot of the avatar-hosts from HIGHRISE: Universe Within, *a web documentary at the NFB (2015), which as co-created with scholars, high-rise residents, and creative technologists. Used with permission from the National Film Board of Canada.*

think the state, or a corporation is going to build out the system in these areas, because often it's not financially beneficial. . . . A lot of people who make claims that the world is going to be flattened by digital technologies simply haven't gone to rural disenfranchised communities to see what the material conditions are like and to listen to the voices of people who live there."

Collective Wisdom coauthor Katerina Cizek co-created a comparable cross-disciplinary project under the NFB's HIGHRISE umbrella, with the digital documentary *HIGHRISE: Universe Within* (2015). She partnered with two academics at the University of Toronto, the critical geographer Deborah Cowen and Emily Paradis, a social work scholar specializing in participatory action research methods. The three teamed up with a group of fourteen residents in two high-rises in suburban Toronto, where Cizek had been working for several years, to develop and conduct a participatory-action survey. The residents asked their neighbors (in more than a dozen different languages) about their digital lives, especially their transnational connections. Members of the team also traveled to a suburban high-rise near Mumbai, India, to ask similar questions. Through these processes, they created media tools for use by the buildings' tenants associations and collaborated on the *HIGHRISE: Universe Within* webdoc project with the digital agency Secret Location as well as on a new software developed in New York City called DepthKit. The tech and editorial teams together developed and iterated on the first documentary use of volumetric filmmaking (which involves the mapping of video images of people onto 3D models to create 3D objects directly in web browsers). Cowen and Paradis et al. also coedited a related collection of essays, *Digital Lives in the Global City: Contesting Infrastructures (2020),* which was presented in an unconventional polyphonic and highly visual academic book. Drawing on documentary processes, critical geography, field research, and technological innovation, the teams co-created new frameworks for understanding high-rise landscapes as well as new forms of tech and media as well as social justice movement building.

The crossover between art and science is growing, and with emerging media such as virtual reality (VR) and AI, the complex connections become almost impossible to untangle. In a 2021 production of *Midsummer Night's Dream* called *Dream*, the Royal Shakespeare Company explored how audiences could experience and interact with live performance remotely online, using motion capture and virtual production. The effort came out of an unprecedented consortium of fifteen UK organizations called the Audiences of the Future Challenge.

JOURNALISM AND SCIENCE MEET

Journalists now also work across disciplines and sometimes use co-creative approaches to address the biggest issues of our times, stories too complex for one reporter, one newsroom, or the legacy journalistic tools of the twentieth century.

The ISeeChange platform, as mentioned earlier, is an example of how co-creative and cross-disciplinary methods are shared between reporters, scientists, and on-the-ground community members. ISeeChange has joined with the National Aeronautics and Space Administration (NASA) "to create a citizen science corps that will correlate community experiences to space-based observations of atmospheric carbon-dioxide levels, season-to-season, year-to-year," according to its website.[6] This transdisciplinary approach comes from the ground up and the sky down, figuratively and literally.

In ISeeChange's first project tracking drought in Colorado (2015), the group's CEO, Julia Kumari Drapkin, stated that "we found that the community flagged environmental trends up to three months in advance of official reports." As she reported

Participants engage with ISeeChange, a platform and global community that uses co-creative and cross-disciplinary methods between journalists, scientists, and on-the-ground community members. Used with permission from Julia Kumari Drapkin.

The web interface of ISeeChange. Photo courtesy of ISeeChange.

in our interview, "So we were talking about drought and wildfire in March and April before Colorado burnt down that fall and summer. We had four trends that we ended up reporting ahead of time, but there were these two insights: one, that residents were experts in their own backyard and we could learn from them and take our cues from them and respond and let them initiate the process, and, two, that this form of record keeping with stories and data and math together can be really powerful for documenting climate change."

Whether tracking rising urban heat in public housing in Harlem or flooding in New Orleans, ISeeChange considers community members to be equal partners in the transdisciplinary approach to documentation and campaigning. As Drapkin stated in our interview, "I come at it from the documentary space, the filmmaking space, radio space—I've worked in all of those spaces. But where do we begin and end, if it's art, if it's science, if it's data, if it's engineering or design? We're just laser-focused on empowering the community to include their voice on what to monitor and how to adapt. . . . Both journalism and science have pushed way too far in distancing ourselves—even failing to recognize that we ourselves are community members."

RESIDENCIES, HACKATHONS, LABS, AND INCUBATORS

Many divergent organizations have sought to cross-pollinate science, arts, and media by creating long-term or temporary subspaces and events to bring practitioners together.

RESIDENCIES

The American arts residency generally follows one of two traditions: in the woods or in the lab. The in-the-woods model has a Walden-like quality that favors the single author. It isolates artists, encouraging them to work in natural, secluded environments perhaps with a cohort of other artists to meet for dinner and discussion. The in-the-lab model, by contrast, throws artists into the contained environment of a scientific lab or the routines of everyday life in government or industry for an extended period. During the COVID-19 pandemic, many residencies shifted to "in the Zoom" playing out entirely online.

Residency models have a spotty but long history. In the mid-1960s the London artist Barbara Steveni launched the Artist Placement Group, or A.P.G. (later Organisation and Imagination, or O+I), with her husband John Latham. Participating artists included Joseph Beuys, Yoko Ono, and other members of the Fluxus group as well as David Toop, Hugh Davies, and more. The purpose was to facilitate residencies for artists at nongallery institutions such as the British Steel company and the UK Department for Health and Social Security.[7] Their resultant work is now held at the Tate Archives. At around the same time in the United States, MIT began to cross modern art with science at the Center for Advanced Visual Studies, founded by Gyrögy Kepes. Meanwhile, artists Robert Rauschenberg and Robert Whitman cofounded E.A.T. with the engineers Billy Klüver and Fred Waldhauer to facilitate projects using video, sonar, optical effects, and other new technologies for art performances and installations. The influential journal for transdisciplinary art *Leonardo* was launched in1968.

"In the artist-technologist collaborations that I've looked at from the 1950s and '60s, the work that went on was primarily ideological," Stanford communications scholar Fred Turner wrote. "It made it possible for engineers who were building our media and communication systems, the Bell Labs sound system, or the engineers working at NASA on rocket engines that would send things into space, or people working in Silicon Valley on Polaris missiles, to imagine themselves as the same kind of exquisitely sensitive and culturally elite person that, say, a John Cage was, or Robert Rauschenberg was."[8]

In his book *From Counterculture to Cyberculture*, Turner argues that the counterculture movements in and around San Francisco created the estuary for the spawning of
Silicon Valley itself.[9] The artistic environment created the conditions for technological industrial boom.

While the in-the-lab models fell out of favor for a period of time, they are making
a comeback in the twenty-first century. Perhaps the most famous example was performance artist and musician Laurie Anderson's three years at NASA in the early 2000s,
which catapulted the trend into public consciousness and grounded space exploration
in poetry. There are now over one hundred such residency programs in the United
States alone, and in 2015 the European Commission launched a major initiative called
S+T+ARTS, which runs residencies, prizes, and regional hubs and in March 2020 in
Paris hosted the first ever international symposium on art and science residencies.

Gerfried Stocker, the artistic director of Ars Electronica, a think tank and festival
founded in 1979 to champion the arts and sciences in Linz, Austria, was quoted in a
2017 *New York Times* interview saying that he "believes that artists have become cultural missionaries in a time of intensive transformation driven by new technology.
It's crucial," he said, "that humanistic voices address the ethical and moral questions
created by this transformation."[10]

But the ongoing tensions of economic imbalances have remixed the models with
unequal and uneven results. As Vanessa Chang, a writer and curator who has researched
the resurgence of experiments in art, tech, and society, says:

> Like the gig economy, these models of collaboration are built on the labor of the creative
> precariat. In their lionization of flexible labor in temporally bounded projects, many of
> these labs operate within an extractive economy. Within the context of cultural produc-
> tion, a logic of extraction understands artistic labor as a resource whose value is to be
> milked and fed into industry for profit. Media artists working with technology firms have
> often found their intellectual property deployed in commercial products—sometimes with-
> out permission or credit. Such an approach does not engage in the labor of renewal or
> regeneration—extraction's opposite—and in creative fields can exhaust artists, organiza-
> tions, and communities pouring their sweat equity into this work.[11]

In that context, the residency model of the modern artist working within a scientific institution still suggests inspiration, with at best some collaboration rather than
true co-creation. More recently, Nokia Bell Labs, involved in early E.A.T. programs,
revived the programs with co-creative but short-lived results. Through their 2017 yearlong artist residency, the dance troupe Hammerstep began experimenting with an
emergent drone technology then in development at the lab. The dancers found the

gesture-control interface for the drones very limiting and through sustained dialogue and experimentation, according to Nokia Bell Labs' Domhnaill Hernon, transformed it:

The co-creative process resulted not only in a public dance performance but [also] significant design changes to the drones' gesture-control algorithms, related wearable devices, and the platform-networking systems within which they operate. [The dancers] opened up the mindset of the engineers about how, through our design of those algorithms, we had really severely limited the creative potential of the technology we were using. . . . They now have a level of adaptability and flexibility that's much more intuitive and natural for controlling the drones that they would not have had if they were left to their own with their research and [were] not engaged with the artistic group.

Other companies such as Facebook, Adobe, Planet Lab, and Autodesk have programs that often bring established artists into the lab for short residencies—often within

Jess, shown at the opening of the installation Street Health Stories, *is one of the artists and co-creators. It served part of the NFB's program* Filmmaker-in-Residence. *In the project, young parents of no fixed address peer-interviewed people experiencing homelessness about their health and created photographic portraits. The resulting work included a contribution to a major report on health and homelessness, a film shown on national television, and an installation that traveled across Canada. Used with permission from the National Film Board of Canada.*

limited scopes—to beta-test tech, find new uses for the products, and also help discover and solve bugs and interface issues.

In another more socially engaged co-creative model, Cizek's five-year residency at a Toronto inner-city hospital with the NFB's *Filmmaker-in-Residence* program included media projects that were framed as research projects in partnership with doctors, nurses, and other community stakeholders. All of these projects were passed through the hospital's ethical review board. The results affected policies at the hospital, in the city, and even at the provincial level, including sensitivity training for emergency and delivery room staff as well as police protocols for dealing with people in mental health crises.

Fellowships also often loosely follow the residency model. The Mozilla Foundation, funded by the Knight Foundation, placed five cohorts of research fellows for ten months each in newsrooms across the United States and Germany to infuse journalism with open-source expertise. More recently, Mozilla has put the journalism fellowships on hold in order to run a variation of the program in which artists, coders, and activists are embedded in international nongovernmental organizations to work on independent research and project development. All of these projects, however, are at risk of losing impact once the residency or fellowship is over.

HACKATHONS, LABS, AND POP-UPS

Residency programs imply a longer-term stay, often with big, established names. Many organizations have tried much shorter, leaner, and faster tests for cross-pollinations as digital culture introduced a new era of collaborative possibilities. One early model was the (in)famous hackathon.

Hackathons began sprouting up in the tech world near the turn of the twenty-first century as a means to pool coding talent and attack a specific problem in a software project while participants we holed up in the same room, often over twenty-four or thirty-six hours. A stereotype emerged from these hackathons: young, White, middle-class coders pulling all-nighters fueled by energy drinks to solve a bug in the system. Soon hackathons even became recruitment tools for companies, poaching talent that had proved themselves at exploitative hackathon-style auditions. Still, the core principle was generative: assemble disparate talent to solve a problem quickly and then come out with a prototype or something that will showcase the idea.

Within a few years, cultural provocateurs began adapting the hackathon model to the world of storytelling. Small groups and think tanks spread the word. The

Junction is the largest hackathon in Europe, with up to 1,500 participants working over forty-eight hours at the annual competition. Corporate sponsorships include Uber and other major companies. Photo by Vmuru. Used under Creative Commons.

Banff Centre for Arts and Creativity's New Media Institute, a pioneer in emerging tech, convened groups early in the 2000s. San Francisco's Bay Area Video Collection residency program nurtured projects such as *Question Bridge* (see the Spotlight in chapter 2) as the first public new media lab in the United States. These hackathon-minded evangelists began to travel. Documentary maker Peter Wintonick roamed the world to festivals with his group Docagora and professed the possibilities of the open web and a connected world in the context of documentary work.

In 2008, *The Guardian* newspaper in the United Kingdom was one of the first journalistic organizations to experiment with the hack model, bringing together "code monkeys with word monkeys" for a two-day hackathon in what has become an annual event at the organization called Hack Days. Through cycles of rapid idea generation and prototyping, technical and editorial staff have together over the years developed key innovations in data journalism such as the 2011 Ophan user analysis tool, which helps editors track in real time the details on uptake and readership of specific stories across the platform. Annual Hack Days have also resulted in fun conceptual projects, such as the creation of a tool for the two hundredth anniversary issue of *The Guardian*, to digitally reformat current news in the typeset of the first issue in 1821.

Installation view of Question Bridge. *Used with permission from Kamal Sinclair.*

Also in London, Power to the Pixel brought together European players from gaming, interactive, documentary, fiction, and other industries to co-create. "It was kind of exciting and, in the ten years, seeing relationships being formed," Liz Rosenthal of Power to the Pixel told us. "People who are usually stuck in their organizations or in their industry silo, coming together and going 'Oh my gosh, we've shared exactly the same issues and are looking at the same challenges and the same creative practices.'"

CrossOver Labs in the United Kingdom also germinated the seeds of rapid prototyping and facilitated fifty or more labs around the world. These labs hired technologists specializing in gaming, open-source coding, and interface design to work with visual artists; story specialists in documentary, feature film, and editing; and specialists from fields as diverse as architecture, agriculture, physics, and botany. Putting such diverse disciplines together in a room came with consequences. Former Cross-Over director Heather Croall, now the director of the Adelaide Fringe Festival, warned that the process of self-examination and creative pressure is tough. "Everyone hates it when they're there, but it's really good to go through it because then that's where you discover really new things," she told us. "In order to deal with that, we used to take people to really beautiful hotels in the countryside that they wouldn't want to leave. It's a bit like going to the [birthing] center or something. But then by the end of the time, it's great."

By 2011 the Tribeca Institute, with its New Media Fund, began holding hackathons that eventually extended into Story Labs to incubate emerging media projects. One outcome was *Do Not Track* by Brett Gaylor, a six-part online "personalized documentary" that accessed users' own data to show how a film and the web can watch you back. Tribeca ran hackathons across the United States and internationally at CERN in Switzerland, and in Germany, France, and Israel. Opeyemi Olukemi, who worked with Tribeca at the time, said in our interview that "when people come together, everyone has a different philosophy of how to work. So, I start from ground zero and I have them create the best way forward. And I don't push. . . . It's more about saying, what can we create together? . . . [It] goes beyond an experience, it goes beyond a project, it goes beyond an event. It's also the ability to really expand your mind to work in a different capacity. And also hopefully build relationships that can also extend in a more impactful way."

New Frontier at Sundance was perhaps the best funded and most influential cross-disciplinary lab currently running in emergent media. New Frontier took the long-standing film lab model at Sundance and morphed it into an annual cross- and beyond-disciplinary week, with a carefully curated selection of six projects organized loosely with a theme. A group of forty people, including filmmakers, technology advisers, and industry mentors, convened at the Sundance Resort. Kamal Sinclair, former director of New Frontier, told us that she always welcomed participants by assuring everyone that they were meant to be there, since many of them privately confessed to her a case of "impostor's syndrome" when they first arrived, given the varied expertise in the room.

Another now defunct Sundance program, called Stories of Change, ran in partnership with the Skoll Foundation, partnering social entrepreneurs with storytellers. The relationships varied from simple coordination of film launches to deep co-creation in which a project was conceived and emerged from the partnerships. New Frontier also had the Indigenous Lab Fellowship, specifically dedicated to supporting Indigenous filmmakers, and has partnered with the University of Santa Cruz's World Building Institute to hold workshops imagining the future of Los Angeles with designers, documentarians, and technologists.

For their part, the London- and New York–based Doc Society created Good Pitch, which developed into an international network of daylong gatherings to help match social and political documentarians with nonprofits, foundations, and

A still from Nairobi Berries, *a VR experience created by Kenyan artist Ng'endo Mukii as part of Electric South. Used with permission from Electric South.*

nongovernmental organizations in an effort to co-create outreach and engagement strategies.

Meanwhile, new kinds of pop-up labs have begun to emerge. "These different models do different things," said Ingrid Kopp, formerly of the Tribeca Institute and now cofounder of the South African nonprofit Electric South. "I've always loved this idea of 'temporary autonomous zones.' You have these pop-up zones, and they can't be commodified or corrupted too much because they don't last for long enough

The 2018 cohort of New Dimensions Lab at Electric South. Used with permission from Electric South.

for any of that to happen. You can't even have someone write a snarky tweet about them. I think there's something really amazing about those."

Electric South's VR+ labs tap Africa's talent in fashion design, music, and photography in an effort to bring more disciplines into immersive media. Wendy Levy of the Alliance for Media Arts and Culture has run Hatchlabs in multiple US locations to partner storytellers with scientists and community organizations in quick match-making sessions. University of Southern California communications scholar Henry Jenkins runs adjacent workshops called Civic Imagination that have a youth and civic focus. The point of these sessions is to bring together diverse stakeholders in a local community, many of whom have been segregated from each other in their daily lives, to imagine their common futures.

Pop-up hacks, labs, and workshops may vary in length, approach, and commitment, but they are all intended to harness the energy that is sparked when people from different disciplines and walks of life are put together. "You have to be really intentional about what it is that those [events] are trying to do," said Kopp. "Hackathons are terrible to build a project, but I think they're great to get people working

The Alliance for Media Arts + Culture's National Youth Media Summit Chicago (2017) featured young media artists, poets, and community activists in collaboration and performance. Used with permission from the Alliance for Media Arts + Culture.

The Alliance for Media Arts + Culture at the OMI Gallery–Community Artist talk for Oakland Fence Project (2017). Photo courtesy of Wendy Levy. Used with permission from the Alliance for Media Arts + Culture.

together to think about different ways of working. It's nice to have a prototype. . . . It doesn't have to go anywhere after that."

CULTURAL INCUBATORS

Beyond the hackathon model, Silicon Valley birthed another method of tech development, the incubator. The most famous is Y Combinator, which was launched in 2005 to provide seed money, space, mentors, and networking within the Silicon Valley ecosystem. Each year Y Combinator accepts two batches of projects and to date has helped over one thousand companies form, representing a combined valuation of over $80 billion. (Y Combinator takes 7 percent equity up front.) Similar tech start-up incubators have mushroomed around the planet.

The bottom line in this context is financial profit, but what if the model could be adapted to stand for something else? New York's New Museum, for example, prides itself on having begun the first museum-led cultural incubator, NEW INC. Former director Stephanie Pereira explained to us that "the museum's incubator is predicated

The VR lab at NEW INC., a residency and incubator space at the New Museum (2019) in New York City. Used with permission from NEW INC.

on the idea of artists and designers and people working in this weird, squishy space in between technology, architecture, civic design, etc. The ideas, the values, the belief system can be incubated by creating a space where those things can come together. We're going to advance culture and also [make] a space that's equitable."

Prior to New Museum, two universities in Bristol in the United Kingdom were modeling a similar framework, collaborating with the media center Watershed to create the Pervasive Media Studio and its multiyear REACT project. REACT was a four-year project in which artists, makers, researchers, creative technologists, and others came together in a series of five initiatives to explore themes such as play, heritage, and documentary.[12] The latter was the setting in which the *Quipu* team developed its work with collaborators in Peru (discussed in chapter 2). The Watershed Sandbox how-to guide details how the methodology enabled "creative risks within a carefully curated community, providing individuals and small companies with space, money and time to work on their most exciting ideas."[13]

In 2020, in-person programs (from residencies to labs to incubators) around the world ground to a halt and pivoted to virtual events in response to the COVID-19 pandemic's sudden constraints on mobility. Some in-person programs were reconfigured to support artists *in situ* within their home communities, as at the Banff Centre for Arts and Creativity. Others went completely online, such as Sundance's New Frontier. Now, the participants met online weekly over several months rather than in the long-standing five-day in-person format. On a wider scale, Sundance has leaned heavily into its co//ab web platform to support artistic communities with resources, meetings, and online spaces for co-creation.

THE POWER OF PERMANENCE

Improvisational and informal project-based partnerships are crucial to co-creation, but there are cases to be made for long-term investments as well. "When projects were finished and teams disbanded, the learnings weren't consolidated," according to Mandy Rose, who runs the Digital Cultures Research Centre, based in the Pervasive Media Studio, and represents UWE Bristol, one of the two universities in that collaboration. "The studio community and the physical space provide important continuity in which ideas and approaches can develop through different iterations across time. As academics, we're part of that dialogue, and research feeds into it and grows out of it."

These types of more permanent hubs are popping up at American universities in journalism, film, and digital culture programs, including Alex McDowell's World Building Lab at the University of Santa Cruz, Lance Weiler's Digital Storytelling Lab at

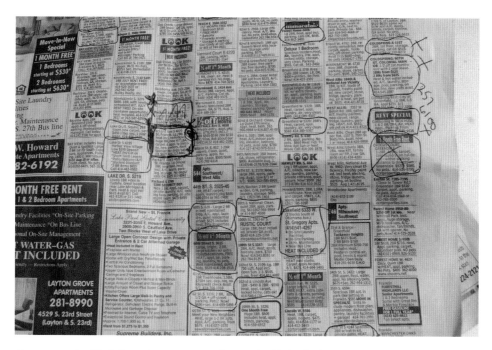

Newspaper classifieds featuring apartments for rent highlight the housing crisis in the United States. Eviction Lab is a co-created, transdisciplinary project that draws on the collective expertise of sociologists, statisticians, economists, journalists, web engineers, and community members who all are engaged in documenting the rising crisis of evictions across America in real time. Used with permission from James Minton.

Columbia University, and Gabo Arora's Immersive Lab at Johns Hopkins. But not all long-term collaborative spaces are academically sponsored. Since 1998, the Eyebeam Center in New York has offered a locale where, according to its website, participants may "think creatively and critically about how technology was transforming our society."[14] Eyebeam provides residents time, space, and money to envision and produce in an open-source environment, as does NEW INC, the new media incubator at the New Museum.

The US-based Guild of Future Architects, launched in late 2019, is another kind of model. Membership-based, the guild strives to bring together hundreds of future-oriented investors, philanthropists, designers, activists, technologists, policy wonks, artists, entrepreneurs, researchers, educators, systems experts, farmers, economists, environmental scientists, and more. According to the executive director, Kamal Sinclair, the guild "aims to raise our collective consciousness for radical transformation, give birth to more diverse and sophisticated organizing forces, and usher in a new era of equitable societies bound by shared values."

In 2018, the DocLab at the International Documentary Festival Amsterdam initiated a five-year research collaboration with MIT's Open Documentary Lab as part of a larger research and development fund. The goal is to shorten the cycle between research and production by linking research findings to project incubation. Both the Open Documentary Lab and the International Documentary Festival Amsterdam's Immersive Network Research & Development Program draw on their complementary strengths and constituencies, pulling together researchers, makers, technologists, and publics to define and explore areas of mutual interest in the documentary and new media space. As a result, the Open Documentary Lab has ready access to the festival's artists and publics, and the International Documentary Festival Amsterdam's DocLab has direct access to MIT's researchers, fellows, and artists, each helping to shape and support the other's agenda and development process.

Also at MIT, the Center for Arts, Science and Technology has since 2012 placed over ninety artists in labs across the school to cross-pollinate and co-create. The center's executive director, Leila Kinney, looks for matches where the influence can flow both ways; for example, the MIT Nano Lab was challenged by an artist to make colors at nano scale, and this turned into scientific research and the discovery of developing color for such a minute scale.

Research and development departments at legacy media institutions come and go, but they do offer a level of in-house stability that can be very healthy for co-creation. The *New York Times* and *The Guardian* each have a research and development lab, as do many public broadcasters around the world. One lab may deal with photogrammetry or augmented reality, another with machine learning, and another with audience engagement. These institutions can foster cross-pollination by putting together unusual teams from different legacy departments. But the teams are often not long in existence, and the perspective can be short. The question to ponder is how innovation and co-creation might lodge more stably in the structures of such organizations. Media scholar Ethan Zuckerman is excited about new employee positions and programs at tech companies that invest more in the long term in interdisciplinary co-creative methods. He lists prominent ethnographers now working for tech companies, such as Genevieve Bell, a senior vice present at Intel, and dana boyd, a principal researcher at Microsoft.

Domhnaill Hernon at Nokia Bell Labs advocates for a deeper cross-disciplinary education, starting in the elementary and high schools:

[In universities, in] the STEM subjects in particular, there's practically zero exposure to the creative side of the world at all. I mean, you're just forced to learn your algorithms and help solve technical problems. And in the actual workplace . . . it's not just a matter of bringing in one artist for a month and then hoping everything will magically be better. That's just a load of nonsense. . . . The creative world does not understand how technologists think and operate, they don't understand how tech companies think and operate, so there's a big disconnect between these two worlds . . . and there's not enough people educating across both sides to bridge those gaps. So, I think that's the biggest challenge, but it's also the biggest opportunity.

In her 2020 research into art and tech in Silicon Valley, Vanessa Chang suggests that "the art and technology field, like the technology sector, is at a crossroads. There has been a surge of interest in art and technology practice from actors who have heretofore invested their time and resources elsewhere—from the art world to global government offices. Artists and cultural collectives in this space are in urgent need of more resilient infrastructures that can support their creative ambitions."[15] She also insists on the need to create neutral organizations and networks to nurture this work outside of corporate and academic paradigms with their own independent governance structures.

OCCUPYING THE THIRD SPACE

Many co-creative practitioners and thinkers have moved so far beyond their disciplines that old definitions don't apply anymore. When we asked practitioners to describe how they identify themselves, one prominent artist simply said "Person." Another interviewee said "Tired," referring to Hannah Gadsby's hit Netflix standup special *Nanette*, which was in part about the performer's weariness of laboring under the label "lesbian comedian."

A few interviewees discussed working in an in-between space, including Lance Weiler, a speculative designer. According to Julia Kumari Drapkin, head of ISeeChange, "We have actually decided to just stop calling ourselves by any traditional label because it doesn't service what we do. We are very transdisciplinary, and we have just been shedding identity right and left in order to get it done, in order to do it right." Danish cognitive scientist Kristian Moltke Martiny describes it as working in a Third Space.[16]

Some funding agencies are also moving beyond siloed approaches. In the United Kingdom, the Audiences of the Future Consortium supported the 2021 production *Dream* (mentioned above) by bringing together the Royal Shakespeare Company with other legacy institutions such as the Philharmonia, the University of Portsmouth, and the new tech company Marshmallow Laser Feast. In Canada, a publicly supported

A former tenant's personal belongings and furniture set out on the curb as garbage. It's an increasingly common occurrence across United States as evictions rise, documented by the Eviction Lab (2017). Used with permission from James Minton.

fund called New Frontiers is a joint effort of three agencies that have historically been separate councils for health, the natural sciences, and the humanities. New Frontiers aims to support international high-risk, fast-breaking research (including media production); in its description, the fund mentions "co-creation of knowledge" but not define the phrase in its glossary. Whether the evaluation and metrics of the review process can accommodate the subtleties of co-creation while still answering to the results-driven expectations of tech research and innovation remains to be seen.[17]

At the MIT Media Lab, every job posting for new faculty lists "antidisciplinary" as a requirement for all candidates. As former director Joi Ito explained on his blog in 2014, "Interdisciplinary work is when people from different disciplines work together. An antidisciplinary project isn't a sum of a bunch of disciplines but something entirely new. . . . Only come to the Media Lab if there is nowhere else where you could do what you want to do. We are the home of the misfits—the antidisciplinarians."[18]

The civic media scholar Ethan Zuckerman said in our interview that dealing with engineers can be an "uphill battle," that is, "to get people thinking culturally about

A hologram of Joi Ito (2016), former director of the MIT Media Lab. Used with permission from Katerina Cizek.

this idea that they might want to work with communities. . . . Whether it's getting people to do ethnography before they design or whether it's more engaged methods like bringing [in] the people that you're hoping are going to use your tech . . . those are still pretty radical where I am."

In Ito's blog, he states that "we will still need disciplines, but I think that it's time we focus on a higher mission and the changes needed in academia and research funding to allow more people to work in the wide-open white space between disciplines—the antidisciplinary space."

In our research of co-creation across disciplines, we've noted an increasing call for recognition and support of the relatively rare unicorns, or "translators," who work in and beyond multiple disciplines (e.g., the unicorn who has a PhD in coding and has exhibited at the Museum of Modern Art). But there is also an urgency to create third spaces for *the many, the masses,* to learn and develop the art and science of collaboration and co-creation. By this, we mean a broad Third Space for the people in which they learn how to work with wider networks, not just another rarefied silo for the exceptional multitalented and multieducated antidisciplinarians but instead a space for all of us to learn to share and work together across silos every day.

SPOTLIGHT: HYPHEN-LABS

by SARA RAFSKY

Hyphen-Labs is a virtual laboratory. Members of the group are a diverse and dispersed collection of women of color based around the globe who, according to their website, collaborate on projects at the "intersection of technology, art, science, and the future."[19] But their core mission is more tied to a defining process and a philosophy for cross-disciplinary practice in emergent media. According to the group's three cofounders—the Turkish-born Barcelona-based Ece Tankal, Mexican Cuban American Carmen Aguilar y Wedge, and Brooklyn-based Ashley Baccus-Clark—Hyphen-Labs was created out of a shared conviction that women of color were not being fairly represented, if at all, in the digital environment. Tankal and Aguilar y Wedge started the group in 2014, with Baccus-Clark joining two years later, in hopes that they could bring together women from different disciplines who shared their concern.

A few of the products developed for NeuroSpeculative AfroFeminism *by Hyphen-Labs, an interdisciplinary studio that creates media at the "intersection of technology, art, science, and the future." Used with permission from Hyphen-Labs.*

The three founders have backgrounds in architecture, engineering, and molecular biology, and their nine international collaborators work in technology, architecture, design, science, and art. Hyphen-Lab's work now encompasses filmmaking as well as product and speculative design.

Collaboration has been an intrinsic part of the work since the group's inception, as has the pooling of their collective interests and skill sets. "I really like oral communication. Ashley is an amazing writer, and Ece is an incredible visual communicator," said Aguilar y Wedge. "Having those three elements together really helps bring a story to life, bring a project to life. And I don't think one can work without the other." The women explained that typically the three of them will develop an idea together and then work with artists, designers, and scientists to actualize it. Since Hyphen-Labs was cash-strapped at the beginning, by necessity the founders collaborated with people who wanted to work on a project solely because they felt passionately about an idea. "I think that's where collaboration has to start," Baccus-Clark said. "Everybody has to love the idea."

The most elaborate project yet to come out of Hyphen-Labs is *NeuroSpeculative AfroFeminism*, which was featured at the 2017 Sundance, SXSW, and Tribeca film festivals. As described on the Hyphen-Labs website, *NeuroSpeculative AfroFeminism* is a "three-part digital narrative that sits at the intersection of product design, virtual reality, and neuroscience, inspired by the lack of multidimensional representations of Black women in technology."[20] In the first part of the *NeuroSpeculative AfroFeminism* experience, viewers enter an installation space that resembles a futuristic beauty salon. The salon is stocked with products that include prototypes for a sunscreen specifically made for dark skin, a camouflage scarf that subverts facial recognition software, a popular earring design repurposed with audio- and video-recording capabilities to use in dangerous situations, and a protective visor that allows the wearer to see out but blocks people from seeing in.

The second part of the project is a room-scale VR experience that places the user in a futuristic neurocosmetology lab. Since Black salons are sites for socializing and political discussion as much as for beauty treatments, users sit in a salon chair and swivel to the mirror to find that they have embodied the avatar of a young Black girl who is set to receive brain-stimulating electrical currents interwoven in her hair extensions.[21] After receiving these "Octavia Electrodes," named in honor of science fiction writer Octavia Butler, the user is transported to a hallucinatory tour of a psychedelic Afro-futurist space landscape, as the team describes in an interview at Docubase.[22] As a third component, the project gathers data to see if the rare use of a Black female

In the first part of the NeuroSpeculative AfroFeminism *experience (2018), viewers enter an installation space that resembles a futuristic beauty salon. Used with permission from Hyphen-Labs.*

protagonist in a VR experience has had any lasting neurological and physiological impact on participants.

As fantastical as the experience of *NeuroSpeculative AfroFeminism* was, each element was drawn from real life. The use of the hair extensions as a plot device, for example, nods to the difficulty of fitting the VR Oculus headset, or the brain sensors used in neuroscience tests, over big hair. The idea of sunscreen for people of color came from an incident in which Baccus-Clark, who is Black, put on sunscreen at the art space Storm King and found that it left a layer of white film on her face. The three women decided to start with speculative products because "so often we found that we either had to modify products to fit us or we couldn't use them," according to Tankal. At a talk at MIT during their time in 2018 as visiting artists at the Open Documentary Lab, a member of the team said that "we wanted to come at it from the aspect of designers, like what are our needs? What can we create that can highlight those needs?"

To realize *NeuroSpeculative AfroFeminism*, Hyphen-Labs worked with over a dozen people in various fields, including the architect and developer Nitzan Bartov, the privacy-focused artist Adam Harvey, the artist Michelle Cortese, and the fashion label

A still from the second part of NeuroSpeculative AfroFeminism *is a room-scale VR experience that places the user in a futuristic neurocosmetology lab. Used with permission from Hyphen-Labs.*

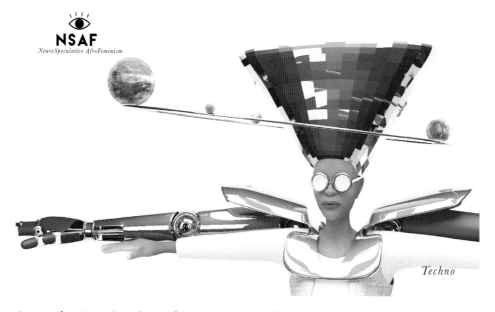

A poster from NeuroSpeculative AfroFeminism. *Used with permission from Hyphen-Labs.*

AB Screenwear. Baccus-Clark said that the group seeks to get away from the overvalued celebration of individualism to focus on communal partnerships; the projects are more successful because the diverse group dynamic forces each member to "push through boundaries in our thinking." Aguilar y Wedge added that they work by "feeding our disparate fields and the jargons of those fields into another and [through] seeing how they coalesce and create new lines of questioning."

Hyphen-Labs also sees its role as co-creating with communities. The lack of accessibility and diversity in design is a central concern, so the group says that Hyphen-Labs looks to bring their projects into communities that have been neglected by the industry. This includes holding workshops and creating projects with high school students as well as with young professionals and specialists, using a methodology based on project-based learning and human-centered design principles. Baccus-Clark said that when they develop projects around themes such as income inequality, criminal justice, and police brutality, they aim to capture "in your experience, in your own body, what does that feel like? And how are you impacted?" In these processes, Baccus-Clark said, there is "no boss-and-subordinate relationship. We want the people we work with to thrive and to feel like they belong in the spaces that we're making, and that their input is valuable." Unsurprisingly, that process isn't always seamless. "Art is inherently personal," Aguilar y Wedge said. "Having another person or people in there trying to mold and shape those lines of inquiry can be challenging." Then there are the constant issues of finding funding. *NeuroSpeculative AfroFeminism*, which received support from Intel, was the group's first funded project after years of the founders using their own resources. The women maintain a steady practice of commercial design work to fund the studio, but for the more artistic and speculative projects they cobble together resources from a variety of sources. Hyphen-Labs continues to prototype some of the speculative products it designed for its futuristic beauty labs—the scarf, the founders said, is the furthest along so far. The trio also has a long wish list of future collaborators that includes artistic luminaries such as Olafur Eliasson, James Turrell, and Bjork, as well as people in philosophy, science, mathematics, and even government. Aguilar y Wedge said she was interested in exploring how one might apply design principles in the latter, particularly at higher levels of office.

The women of Hyphen-Labs revel in the slippery, wide-reaching definition of co-creation, using words such as "poetic," "liquid modernity," and "futuristic" to describe their own concept of the practice. These terms could also be applied to the work of Hyphen-Labs as it strives to push the idea of what design is and who it is for.

SPOTLIGHT: *SGAAWAY K'UUNA (EDGE OF THE KNIFE)*

SGaaway K'uuna (*Edge of the Knife*) is a $1.89-million 2017 feature film shot entirely in the Haida language and owned by the Haida Nation. The film, set in 1830, tells a traditional Haida story of a "Wildman" who through crime, guilt, and abandonment becomes lost and feral in the forest. Through ceremony and healing, he is returned to the fold of his community.

The film was co-created in an unusual three-way partnership between the Council of the Haida Nation, the world-renowned Inuit film production company Isuma, and Leonie Sandercock, a non-Indigenous community and regional planning professor at the University of British Columbia. Together, they accessed an unusual mix of funds that allowed for a co-created and ambitious set of stages and goals for the project, and yet the project remained rooted in the community-driven aspirations of local economic development and Indigenous language revitalization.

"The secrets of who we are are wrapped up in our language," Gwaai Edenshaw, a codirector of the film, told the *New York Times* in a 2017 interview. "It's how we think,

On the set of SGaaway K'uuna (Edge of the Knife) *(2017). Photo by* شون حرف *Farah Nosh. Used with permission from Isuma Distribution International.*

how we label our world around us. It's also a resistance to what was imposed on us."[23] Today, fewer than twenty fluent speakers of Haida are left after over a century of colonialist policies and practices. In more recent decades, though, the Haida Nation has been politically resurgent, mounting some of the strongest political and legal campaigns for Indigenous sovereignty, including an as yet unresolved land claim at the Supreme Court of Canada.[24]

The primary partner and majority owner of the film is the Council of the Haida Nation (via a company called Niijang Xyaalas Productions) and owns all intellectual property and had the power to hire and fire anyone on the project. The council is the governing body of the Haida people who have occupied Haida Gwaii since time immemorial. The traditional territory encompasses parts of southern Alaska, the archipelago of Haida Gwaii (forested islands off the coast of what is now Prince Rupert, Canada), and its surrounding waters. The precontact population is estimated at tens of thousands who lived in over three hundred villages, but according to the Haida Nation's website, "During the time of contact our population fell to about 600, . . . due to introduced disease including measles, typhoid and smallpox. Today,

On the set of SGaaway K'uuna (Edge of the Knife). *Photo credit* شون حرف *Farah Nosh. Used with permission from Isuma Distribution International.*

Haida people make up half of the 5,000 people living on the islands."[25] The Haida Nation now resides in two reserves, known as the villages of Skidegate and Old Massett, in which distinct dialects of the Haida language are spoken. Additional Haida Nation members live in the south in Vancouver and in other urban areas as well as in the north in Haidaberg, Alaska, which has its own dialect.

Team members from Isuma are the second partners on the film, led by executive producer Zacharias Kunuk. Isuma is based over three thousand kilometers away from Haida Gwaii, over land, water, and ice, in the northern Inuit town of Igloolik and produced, among other works, the acclaimed *Atanarjuat: The Fast Runner* (2001), filmed entirely in Inuktitut. The production and creative team of Kunuk and Norman Cohn had been seeking to replicate the Isuma model of storytelling in another Indigenous nation and finally found a partner in the Haida Nation.

"The community is of central importance to this type of filmmaking," the producers said in reference to the Isuma model in one of the early fundraising proposals for *Edge of the Knife*. "Community members are the primary audience. They are future cast and crew members, they are the prop and costume designers, and they are the

On the set of SGaaway K'uuna (Edge of the Knife). *Photo by Graham Richard. Used with permission from Isuma Distribution International.*

storytellers. Storytelling is an ancient Indigenous practice. Within Indigenous communities, there is a host of storytelling talent, and what better way to create a dramatic film script than with local storytelling experts."

STORY-GATHERING WORKSHOPS

The project began not with a formulated idea or story but instead with a yearlong community-engagement process enabled by a 2014 academic research grant from the Social Sciences and Humanities Research Council (SSHRC) through Leonie Sandercock. She had been working with the Skidegate Band Council since 2012 and initially discussed with the council the idea of a documentary. The band brought it to the Council of the Haida Nation, which agreed enthusiastically.

Sandercock's research design allowed for the SSHRC funds to pay Haida Nation members to participate in two story-gathering workshops, one in each village. "We went in with a completely open slate of what we could do," explained Jonathan Frantz, a non-Indigenous producer and director of photography on the film who studied with Sandercock at the University of British Columbia and had also worked with Isuma in Igloolik for several years. "The Haida Nation had some parameters: that it's in the Haida language, that it's roughly balanced between the two communities [in terms of] hiring and people involved in the film, and that we try to hire as many Haida people as we can," Frantz said.

The workshops were based on community and participatory planning models that incorporated "tables of ten people, a facilitator, a note-taker, and we just wrote down as many ideas as we could in a few hours," as Frantz described it. "We had some food and just basically listened to people talk about any parts of a film that they would like to see." Community members quickly identified numerous potential storylines, along with some specific cultural technology and traditions they wished to see onscreen. Two more workshops followed, this time with an emphasis on the process of script writing. A writing competition then helped identify storylines and writers.

The team ultimately had to turn a few story ideas down due to budget constraints—notably, two strong proposals for stories of reincarnation. "Reincarnation is a very important piece of Haida cosmology, and they were wonderful stories," remembered Sandercock. "And I was really passionate about doing a reincarnation story, and so were the writers." But Frantz was already thinking about the budget as a producer; a storyline with reincarnation would have involved three time periods and three sets of locations and costumes.

On the set of SGaaway K'uuna (Edge of the Knife). *Photo by Graham Richard. Used with permission from Isuma Distribution International.*

The group settled on the Gaagiixid/Gagiid, the Haida "Wildman" story, as the backbone to the script and identified three Haida writers, Gwaai and Jaalen Edenshaw and Graham Richard, the winners of the writing contest. The next stage of development involved six months of writing (with Sandercock as mentor), this time funded by a conventional film grant from Telefilm Canada because paying for actual script writing was "outside the purview of the SSHRC grant," said Sandercock.

"There are different times in the film process that it's great to have a lot of voices and people involved," said Frantz, "and then there are times where, just for efficiency and logistics of getting things done, it narrows down into the hands of a few key creative people. . . . I think throughout the whole project that a narrowing of involvement happened. At the end of the day, the script was written by four people."

Nonetheless, throughout the process the group worked closely with an elders' advisory group of fifteen elders from both communities, including Diana Brown, age sixty-nine, a language advocate who started teaching Haida in the schools in the 1970s. As Sandercock et al. wrote in the academic journal *Planning Theory and Practice*, "The writing team worked closely with elders and knowledge holders from both

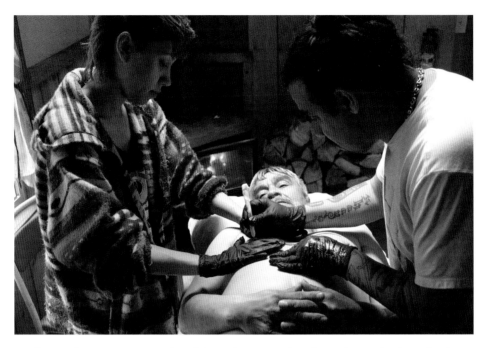

On the set of SGaaway K'uuna (Edge of the Knife) *(2018). Photo by Graham Richard. Used with permission from Isuma Distribution International.*

villages, from whom they sought advice about specific scenes (such as a ceremony to welcome the first salmon caught for the season), and these knowledge holders read each draft of the script and ensured cultural accuracy and sensitivity."[26]

The script was then translated into the two distinct Haida dialects, Xaad Kil and Xaayda Kil, as the Council of the Haida Nation wanted to ensure that both were represented in the film. Finally, producers submitted the finished script to the Canadian Media Fund and successfully financed a line budget for the production.

THE PRODUCTION PROCESS

Preproduction began in 2016 with the hiring of two Indigenous codirectors, Gwaai Edenshaw (Haida), one of the scriptwriters, and Helen Haig-Brown (a member of the Tsilhqot'in Nation living on Haida Gwaii with her Haida partner). Edenshaw and Haig-Brown ran casting workshops and auditions in the two Haida villages, following the Isuma model of working with untrained community members. They recruited the cast and crew from the communities, scouted for locations, and hired costume and props designers, builders, and consulted knowledge-keepers, all of this with a strong emphasis on historical and cultural authenticity.

The process was slow and delayed, with stops and starts resulting from various expectations from funders, partners, and making sure that politically the project was on good footing with all the community participants. A key location, for example, was eventually abandoned because they were unable to gain the right consent and approval through the governing protocols. Before they could begin filming, the cast not only had to brush up on acting and learn their lines but also had to learn how to

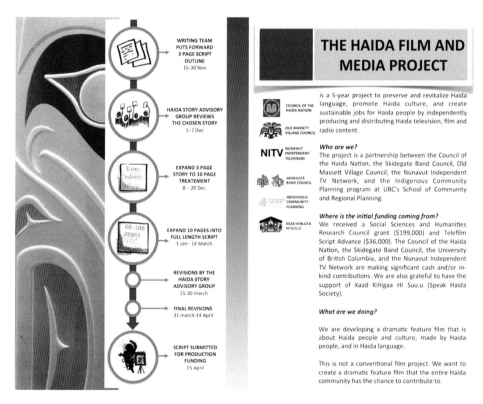

Early flyer (left and right) (2017) created by the three-way partnership to explain the SGaaway K'uuna (Edge of the Knife) *project and process to Haida community members. Used with permission from Isuma Distribution International.*

pronounce the Haida language. The team ran a two-week boot camp focusing on the complex pronunciation, grammar, and structure of the two dialects; Haida contains thirty-five consonants and two tones, and there are twenty sounds that are not present in English.

LONG-TERM BENEFITS

In terms of economic development, Sandercock estimated that over CAD$800,000 went directly into employment in the communities. But that's not where the vision ended. "This was always discussed from the beginning as a pilot project," said Frantz,

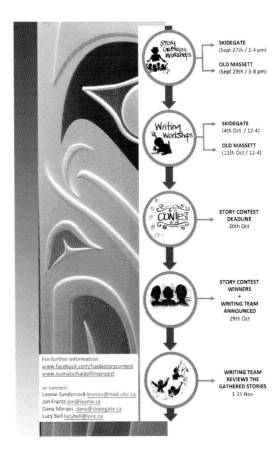

We have a number of objectives for this film:

- Support the active use of Haida language;
- Showcase Haida culture;
- Create employment and business opportunities for Haida people;
- Link Haida people to an international Indigenous media network.

To achieve these objectives we cannot follow a traditional film making approach, which typically follows one person's vision – the director.

Our approach follows the collective vision of our community, from youth to Elders. From the initial concept to the final distribution we will provide opportunities for Haida people to play an active role in the film project.

To support Haida involvement, our project partners will provide mentoring and training opportunities throughout the length of the project. All you require is a keen interest in the project and willingness to learn.

We will also create our own Haida production company that will employ local people and create a media practice that uses film, video and radio to tell stories from the Haida point of view. The possibilities of what we can produce are endless. We hope this initial film project will be the first of many.

Project Timeline
SCRIPT DEVELOPMENT: May 2014 – APRIL 2015
FUNDING APPLICATIONS: April-May 2015
PRE-PRODUCTION: August 2015 – July 2016
PRODUCTION: August 2016 – October 2016
POST PRODUCTION: November 2016 – February 2017
February 2017 – community screening and celebration.
April 2017- Dissemination to film festivals, broadcast on Television, etc.

For further information:
www.facebook.com/haidastorycontest
www.isumatv.haidafilmproject

or contact:
Leonie Sandercock leonies@mail.ubc.ca
Jon Frantz jon@isuma.ca
Dana Moraes dana@skidegate.ca
Lucy Bell lucybell@uvic.ca

Story Gathering Workshops — SKIDEGATE (Sept 27th / 1-4 pm); OLD MASSETT (Sept 29th / 5-8 pm)

Writing Workshops — SKIDEGATE (4th Oct / 12-4); OLD MASSETT (11th Oct / 12-4)

CONTEST — STORY CONTEST DEADLINE 20th Oct

STORY CONTEST WINNERS + WRITING TEAM ANNOUNCED 29th Oct

WRITING TEAM REVIEWS THE GATHERED STORIES 1-15 Nov

"as the start of a potential local industry in the Haida Nation and one of many films to come." Sandercock has since received another SSHRC grant, in partnership with the Haida Nation, to hire local researchers to gather community evidence and analyze the economic and linguistic impact of *SGaaway K'uuna* (*Edge of the Knife*).

"I like the prospect of opening a new window of storytelling for us, and seeing what comes out of that," Edenshaw told the Canadian Broadcasting Corporation in a 2017 interview. "That's what we're looking at—the potential, if we take to this as a nation, the potential for [film] to grow into a full-fledged industry."[27]

4

PLANTS, ANIMALS, GODS, AND AI: CO-CREATING WITH NONHUMAN SYSTEMS

Can humans co-create with nonhuman systems? Increasingly, artists, technologists, and journalists are working with living cells, mold, and ecosystems as well as artificial intelligence (AI) systems and technological infrastructures. But are these really cases of co-creation?

Over the millennia, humans have situated themselves on a continuum of plant, animal, and spiritual realms, as with ancient gods depicted bearing human bodies with animal heads. While the narrative of Western, Eurocentric civilization has often been one of human dominance over nature, even activities such as beekeeping and fruit-tree grafting suggest a level of human respect for and creative interaction with nonhuman agency.

In the more recent history of the development of AI, analogies to biological, evolutionary, and animal systems are pervasive. British cognitive scientist Margaret Boden has mapped the historic intersections of AI innovations with psychology, biology, and the study of social insects. She argues that while the notion of machine intelligence was first floated as a deliberate provocation by Alan Turing in his 1950 manifesto "Computing Machinery and Intelligence," it has since influenced and been influenced by our understandings of the brain, the mind, and other biological and nonhuman systems.[1] Waves of artists and computer scientists have produced media inspired by and in collaboration with biological processes as well as machine systems. Journalists and documentarians have now also begun interrogating the limits, the potential, and the risks of AI technologies.

Collage created by the Helios Design Lab. Photo courtesy of Priya Shakti and Ram Devnani.

Alan Turing at age sixteen in 1927. Turing is one of the earliest theoretical computer scientists and first floated the idea of machine intelligence in his 1950 manifesto "Computing Machinery and Intelligence." Image under Creative Commons.

The prevailing public discourse on the subject dwells on dichotomies such as whether humans will control these systems or AI will "take over." Are they merely tools or something more? Are we headed for Frankenstein or Shangri-La? Are humans on the road to immortality or extinction? Better questions to pose might include how much of this is science fiction, crossing over from speculation to dangerous deception, and how do training sets of data take form as they are fed into machine learning systems. In *Atlas of AI* (2021), Kate Crawford seeks to map out "a theory of AI that accounts for the states and corporations that drive and dominate it, the extractive mining that leaves an imprint on the planet, the mass capture of data, and the profoundly unequal and increasingly exploitative labor practices that sustain it."[2]

So too, for decades artists have been exploring these questions in their work with nonhuman and AI systems. We have interviewed over thirty artists, journalists, curators, and coders specifically to ask about their relationships with the nonhuman systems with which they work. Many trace their lineage to Donna Haraway's *A Cyborg Manifesto* (1985) as well as to technofeminism, posthumanism, and studies of the Anthropocene. Others were inspired by the environmentalist land art movements of the 1960s and 1970s, but not all. Some are computer scientists cowriting baroque music with algorithms; others create sculptures with bees. Some are exploring the boundaries of deepfake and synthetic media technologies. Others are working to expose the underbelly of the technological surveillance systems we are entangled in. Might the notion of co-creation contribute to the meaning of this work, perhaps by modeling alternative and justice-based approaches? And how might artists help expose the gaps in the dominant narratives about these systems?

The possibility of co-creation between humans and nonhuman systems, whether other living beings or the nonliving entities that we today call artificially intelligent, provokes difficult questions regarding the conditions for co-creation as well as the nature of creativity and consciousness. One such question concerns whether co-creation requires equivalent agency. If nothing more, the very question helps us to interrogate our human condition regardless of whether one accepts the possibility of other agencies.

One might argue that the long coevolution of humans and other living organisms, whether in the form of hunting dogs and falcons or intestinal microflora, demonstrates a robust history of interdependencies. Drawing on the creativity of dogs and falcons to help with hunting seems a different order of cooperation than the essential work of microflora in the human gut. But until relatively recently, the issues of animal consciousness—its existence, its gradations, and its scope—have remained largely outside the Western philosophical debate. In her book *The New Breed (2021)*, Kate Darling situates our relationship with robots within the context of our long, deep history of living and working with animals. "For millennia, we've relied on animals to help us do things we couldn't do alone. In using these autonomous, sometimes unpredictable agents, we have not replaced, but rather supplemented, our own relationships and skills."[3]

René Descartes, a foundational figure in seventeenth-century Enlightenment thinking, compared animals to automatons. He asserted that only humans were capable of rational thought and of operating as conscious agents, while animals simply followed the instructions hardwired into their organs. Only at the end of the nineteenth century did a more nuanced view slowly begin to take root as Charles Darwin's notion of mental continuity across species found adherents. Today, although there is no consensus, according to the *Stanford Encyclopedia of Philosophy* "many scientists and philosophers believe that the groundwork has been laid for addressing at least some of the questions about animal consciousness in a philosophically sophisticated yet empirically tractable way."[4]

If at least some thinkers entertain the idea of consciousness in nonhuman living beings, what about machines? If humans are to co-create with machines and not use them simply as tools (the mechanical equivalent of draft animals), then the same problems of consciousness and agency may apply. Here too the verdict is out, but the language that has been used to describe some computational systems has framed the issue in the affirmative. The terms *electronic brain*, *mathematical Frankenstein*, and *wizard* were used to describe the 1946 appearance of ENIAC (Electronic Numerical Integrator and Computer).[5] John McCarthy coined the term *artificial intelligence* in 1955. And in 1958 it was claimed that the "electronic brain" of the "Perceptron" would one day "be able to walk, talk, see, write, reproduce itself and be conscious of its own existence."[6] These visions all harkened back to Ada Lovelace's 1840s proposition that machines "might compose elaborate and scientific pieces of music of any degree of complexity or extent."[7] In these instances, the case for machine consciousness, intelligence, and creativity was asserted rather than argued.

The implications of these assertions have been mixed. Both unrealistic expectations and fears have triggered a cycle of decade-long AI winters in which research funding dropped precipitously and programs closed, only to be revived by the next overhyped innovation. This boom-and-bust pattern remains with us, with MIT in the fall of 2018 announcing plans for a new college focused on AI, backed by $1 billion in funding.[8]

On the other hand, disciplines such as psychology (theories of human knowledge acquisition), neuroscience (including the study of biological self-regulation), and research areas such as the biotracking of insect behavior (framed as "distributed intelligence") have all found relevance in AI research. Descartes's connection between animals and machines, even if intended dismissively, has come full circle. This is an especially salient point at a moment when our dependence on statistically based deep learning in current AI systems seems to be reaching its limits. The reasoning- and knowledge-representation side of AI, inspired by biological systems, may well lead to the next step in machine intelligence.

Musician David Cope rightly asks about the implications that different AI systems have for agency and with them the possibility for creative partnerships. Is machine

Ada Lovelace, English poet and mathematician, in a daguerreotype from 1843 or 1850. Lovelace is considered the first to express the potential of the "computing machine" and the world's first computer programmer. Image under Creative Commons.

learning akin to training a pet, or might AI systems yield more robust forms of creative interaction? Co-creative engagements might offer ways to test and interrogate the various configurations of intelligence and agency that AI systems, animals, plants, and even bacteria might display. The parameters of co-creation—of ethically reframing who creates, how, and why—have newfound relevance to the extent that we grant or even entertain the possibility of intelligence in some of these nonhuman entities.

Haraway's legendary *A Cyborg Manifesto* proposes that cyborgs might transcend gender, disciplines, and species. The posthumanists more broadly decentralize humans on a broader spectrum of species and ecological systems. Indigenous scholar and new

Still from Jason Lewis's video game Otsì! Rise of the Kanien'kehá:ka Legends, *produced in the Skins 1.0 Video Game Workshop at the Kahnawake Survival School, 2009. Used with permission from Jason Lewis and Skawennati.*

media artist Jason Lewis and his collaborators write that Indigenous epistemologies are "much better at respectfully accommodating the non-human."[9] They cite Blackfoot philosopher Leroy Little Bear to suggest that AI may be considered on this spectrum too: "the human brain is a station on the radio dial; parked in one spot, it is deaf to all the other stations . . . the animals, rocks, trees, simultaneously broadcasting across the whole spectrum of sentience."

As humans manufacture more machines with increasing levels of sentient-like behaviors, we must consider how such entities fit within the kinship network and in doing so address the stubborn Enlightenment conceit at the heart of Joi Ito's critique in his *Resisting Reduction* manifesto that humans should prosper by controlling nature rather than flourishing in harmony with it.[10] Co-creation offers a hands-on heuristic to explore the expressive capacities and possible forms of agency in systems that have already been marked as candidates for some form of consciousness. Only by probing those possibilities will we be able to move beyond blanket assertions or denials of agency and interrogate ourselves, critically, in the context of possibly intelligent systems.

LIVING SYSTEMS

The artists we interviewed who work with living systems may not always use the term *co-creation* to describe their work, but most considered it a workable term. These artists acknowledge the fragility of these living systems—often much of their artistic practice involves the struggle to keep the systems alive—and they depend on and respond to the surprises that these systems might afford. For most artists, these living systems are not mere tools, media, or instruments but instead are complex systems that often require humility and patience. Like other types of co-creation, this work is heavily process-driven.

The Canadian artist Agnetha Dyck works with bees to build artistic creations from found objects, such as broken porcelain dolls. "The bees have the skills of an architect," she told CBC TV's *Artspots* in 2006. "It's their ability to construct up, down, [and] in three dimensions that interested me. They create the most beautiful environment that I've ever seen. I mean, it's just absolutely gorgeous. You have to be an artist to be able to do that. We're so meshed in what we do, the bees and I. They work by instinct and I work intuitively. And I'm trying to figure out whether there's a difference."[11]

For Dyck, the artistic work with the bees is also a rallying call for environmental awareness. "I'm really concerned for them—95 per cent of wild honey bees have disappeared. When you're so close to a creature that's so important to the world and you know how quickly they could disappear, and what that would do to humanity, that's a relationship that's pretty precious."[12]

Collective intelligence runs through the oeuvre of Agnieszka Kurant, an artist based in New York City and in residence at MIT in 2018. She has co-created with microbes, slime mold, insects, machines, and people, including thousands of users of Amazon Mechanical Turk, a crowdsourcing web services system owned by Amazon that negotiates the hiring of freelancers to perform tasks, often small amounts of piecework. She is fascinated with emerging scientific discoveries in the behavior of cellular structures. "More recent research indicates that slime mold also has a capability of learning, which is totally incredible," she told us. "How can something devoid of a nervous system and brain learn? But it does. For example, it's learning if it's in an adverse environment—when there are some negative factors, it will not repeat the same action because it remembers. It has some form of memory, collective memory."

Installation view of Agneiszka Kurant's A.A.I 2, 4 & 5 *(2014), termite mounds made from colored sand, gold, glitter, and crystals. Used with permission from Agnieszka Kurant.*

Kurant's work extends the concept of dispersed cognition, a form of intelligence studied by entomologists and scholars of collective memory to humans but also beyond. As she stated in our interview, "There are new questions of how different kinds of traumas, both collective and individual, influence the microbiome of a person, the bacterial composition of a single human. And . . . we already know for certain that the microbiome composition of a human is responsible to a high degree for our mental processes, for our mental state. There's a very strong correlation between the microbiome and depression, for example, or other mental instability that was for centuries assigned to other factors."

Not all cell-wrangling artists perceive the relationship as wholly co-creative. Bioartist Gina Czarnecki would not use the term to describe her interactions with her two daughters' skin cells, which she combines with 3D glass-mask replicas of their teenage faces. Her primary co-creative relationship is with the scientists she works with, she said, and she equates the cells themselves to goldfish that require tending. She did describe her relationship with technologies as a feedback loop: "I think they are mutually entwined, because obviously the technology gives us ideas, and ideas develop the technology." Her core artistic and political concerns lie with the bioethics of technologies that have medical, cosmetic, and life-extending potentials.

Installation view of Heirloom, *Gina Czarnecki and John Hunt (2016). Used with permission from Gina Czarnecki.*

Alternatively, Canadian bioartist WhiteFeather Hunter uses the co-creative model to describe how she works with microorganisms to create often colorful living biotextiles. Hunter said that she prefers the term *co-creation* over *collaboration*, feeling that the latter implies informed consent. "I would call it co-creation because that acknowledges that [the microorganisms] do have some agency in the process, but it's not taken for granted that they're there of their own free will." The artist works with pigment-producing bacteria such as *Serratia marcescens*, which commonly appears as pink slime in showers and hospitals, where it feeds on soap scum. She also grows mammalian cells into sculptural tissue on handwoven textile forms. Hunter spends much of her time nurturing, feeding, and caring for bacteria and mammalian cells, along with the elaborate environments she creates in which to grow and support them. She said in our interview that "I like to insist upon using empathy as a laboratory technique. It's the same way that people who talk to their plants get really beautiful, luscious plants. So, I talk to my microorganisms. I anthropomorphize them."

Like Dyck with her bees, Hunter sees these creations as microcosms. "We're so steeped in philosophies of humans being at the top of the system that we are alienated from all these other systems," she said. "That's what is leading to fear and huge, huge knowledge gaps. I think it's what's caused all of the trouble that we're in on a planetary scale right now as well."

Other living-systems artists work at a planetary scale. Marina Zurkow, an artist and professor at the Tisch School of the Arts at New York University, described the systems she works with as large-scale natural, atmospheric, and geological ones. She focuses on "wicked problems" such as invasive species, superfund sites, and petroleum dependence and has used life science, biomaterials, animation, social art, and software technologies to foster intimate connections between people and nonhuman agents.

In our interview, Zurkow reacted to "co-creation" as a word that is overused in grant-writing proposals and can be tyrannical, as "it puts undue pressure on artists to create situations that may not make great art." She stressed the need for rigor and tight frameworks. Yet over the course of our discussion, she outlined countless features in her process that correlate with co-creative models: her time frames are very long, she runs multiple projects at the same time, and she works with others in a process-focused

Production image of Marina Zurkow's Mesocosm *(Wink Texas, 2012), a software-driven animation. Used with permission from Marina Zurkow.*

approach. Within her themed collections she has many collaborators, including horticulturalists, playwrights, chefs, and scientists. She produces multiple outputs. In one project strand Zurkow has been exploring the ocean, which has led to dealing with the logistics of infrastructures and commodity flow, such as the shipping routes and internet cables that cross the seas.

More recently, Zurkow conducted a prototype edible-art installation/happening in California, with haute cuisine chefs aiming to incorporate an invasive species into their menus: "We've been looking at regional food systems, and . . . the poster child for the project is the jellyfish. That's not regional, particularly, but it's also problematic there. Jellyfish signal anthropogenic changes to oceans." She added that "these are really things that are processes of discovery and conversation and research and experimentation with materials and through all of that." Her work with fungi, for example, has proven to be an "interesting kind of analogy for bigger systems problems around predictability, reliability, stability, expectability. I'm really interested in these other spaces of creating and, I guess, co-creating."

Transspecies ethnography is an emerging field that elevates anthropomorphizing to an entire methodology. Helen Pritchard, a speculative artist and scholar, has used ethnographic methods to trace animal behavior and other kinds of entanglement. In her project *Animal Hacker*, she has examined how sheep manage to escape the gaze of surveillance cameras at a river crossing. She has also written algorithms that generate poetry from sensing the behavior of algae and has studied the life of a Hong Kong internet-star cat called Brother Cream that attracts visits from hundreds of tourists each day. "There's a lot of speculation that [if humans disappear], the cat will continue as a successful species, and that cats have always exploited human infrastructures in order to become a successful resident of the city. . . . If we change the way we think about cats online, rather than just a kind of cute aesthetic[,] . . . we can perhaps start to think about what other potentials or possibilities there might be for nonhuman animals in the network."

We believe that artists working with living systems can easily relate to the framework of co-creation: they are concerned with the Anthropocene, and they describe their work as process-driven and without a predetermined script conceived by one author.

THE CO-CREATIVE PARADOXES OF AI

We asked artists, journalists, and documentarians who work with AI to describe their relationships with artificial, nonhuman systems. Their answers often reflected a broader spectrum of co-creation, though most also wanted to broaden the social conversation and complicate issues of agency and nonagency, technology and power, for the sake of human and nonhuman futures alike.

VISUAL ART, MUSIC, TEXT, AND ROBOTS

In 2018, Christie's became the first legacy auction house to sell an AI-generated art-work, *Edmond de Belamy*, developed by the French art collective Obvious. It's a portrait produced by a system of two competing AI systems known as a generative adversarial network,[13] which was trained on 15,000 portraits of people painted from between the fourteenth and twentieth centuries in order to create a new work mimicking the characteristics of the human-made portraits.[14] The portrait sold for $432,500, more than forty-five times its presale estimate.

Artists, musicians, and provocateurs have been working with systems that could be broadly described as AI for over half a century. Many makers today are self-taught coders, while earlier waves of artists tended to be university-trained computer scientists who wrote their own bespoke code, drawing on many types of AI systems. Through the open-source movement, makers now are able to access new and easy-to-use, off-the-shelf AI packages that require only basic coding.

In a paper published in 1973 Ernest Edmonds asked, "Will AI eventually replace the artist?"[15] He concluded "No." Over his long career as an artist and professor in computational art, Edmonds has revisited the question many times, always arriving at the same answer. We asked him if his relationship with these systems might be better described as co-creative, and he rejected this model too. "In terms of the computer, I think there's a philosophical objection that I had to the notion of the computer taking over because, of course, that question was posed in relation to art. Art is a human activity for human purpose, for human consumption, consideration, and enrichment. And the making of the art is as much a human process as the consumption of it. And so, I would say that if machines could make art for machines, that would be fine. But it would not necessarily have any relevance whatsoever to human beings."

Edmonds did say, however, that others in his field likely would have favored the co-creation model, including the late Harold Cohen, a pioneer in computer-generated art. "In the last few years of his life, when he took the ideas as far as he could, [Cohen] talked about the computer as his partner. He let the computer program that he had written—a critical point to always remember—generate an artwork, which he then modified. So he used it as almost the work, but not quite the work. . . . I think he's a very interesting example of co-creation."

Algorithmically derived art is a long-standing genre. Roman Verostko and the Algorists were an early 1960s group of visual artists who designed algorithms that generated art and later in the 1980s fractal art. More recently, the Google Deep Dream project reignited public curiosity about AI-generated art and its psychedelic reproductions of patterns within patterns.

Like Edmonds, a new generation artist and photographer, Joseph Ayerle, rejected co-creation as a way to describe his work with AI. He was the first artist to use deepfake technology in a short film, using off-the-shelf AI to map a face from 2D photographs onto 3D-modeled representations of other peoples' bodies. "The software is so incredibly buggy," he told us via email, ". . . and I had so many painful setbacks, that I don't have a good relationship with this tool. The difference to other tools is that I was at the beginning a little suspicious. I worked on a separate PC, not connected with the internet. But bottom line: It is just a tool." AI, he wrote, may "create, but it is not creative."

Many argue that AI experiments in visual art and, to a lesser degree, music have produced more satisfying outcomes than AI texts in the genres of poetry and narrative, which rely on semantic systems. On the other hand, AI-generated poetry need not necessarily have a semantic level in the usual sense. In the hands of scholar and poet Nick Montfort at MIT, for example, AI texts are word art but are not always conventionally meaningful, which is part of a long predigital poetic tradition. However, Simon Colton, a British AI artist and scholar, told us that despite Roland Barthes's proclamation of the "death of the author," most audiences rely for grounding on at least putative backstories of authorship; without that, there is what he calls a "humanity gap."

Kyle McDonald, an AI artist based in Los Angeles, agreed. "I think the big question is less about whether human artists and musicians are obsolete in the face of AI and more about what work we will accept as art or as music. Maybe your favorite singer-songwriter can't be replaced, because you need to know there is a human behind those chords and lyrics for it to 'work.' But when you're dancing to a club hit, you don't need a human behind it—you just need to know that everyone else is dancing too."

Samim Winiger, a Swiss-Iranian designer, artist, and entrepreneur based in Berlin, works with computer-generated text and describes it as computational comedy, which he calls in our interview "a defining human characteristic. [Comedy is] very complex, much more complex than, let's say, painting or visual art, and so forth. . . . There's temporality to it. You need to have historical embedding, sometimes, to understand a joke; you need to have cultural understanding, and so forth. It's a beautiful thing." Winiger said he won't publish his artistic work if it doesn't make him laugh. "I've [computer]-generated 'TED Talks,' because that is inherently already kind of a weird comedic . . . at least for me, it's a very comedic format. You know, [the lecturers] are like machines talking, even though they're humans. I've generated lots of 'TED Talks' . . . like millions of them. They sound reasonable when you read them, but they're completely nonsensical, Dadaistic, machine-generated things—internally coherent and yet completely incoherent. It's a beautiful thing."

In music, David Cope was one of the first to explore the potential of AI, which he turned to after he suffered writer's block in 1980. He built Experiments in Musical Intelligence, which he named "EMI (Emmy)," an AI that mimicked and replicated classical music styles, mostly those of dead composers. In our interview, he told us that he did not describe his relationship with the now-retired EMI as particularly co-creative but would say so of his work with another AI, EMI's daughter, "Emily Howell." He describes their interaction as conversations. He first inputted one thousand pieces created by "her mother" EMI and then interacted with Emily, writing music line by line. "That is, I work with her and she suggests things . . . I tell her what kind of things I want to compose or write or do or whatever, and she has a database just like EMI does, and the material in that, for her, musically, is all of her mother's music."

Anthropomorphizing AI systems is a playful co-creative model adopted by other artists as well, especially those interacting with robots. New York City–based artist

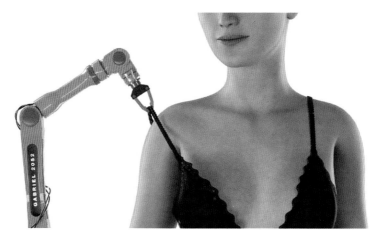

Fei Liu's project Build the Love You Deserve *is a performance series centered on the semifictional relationship between the artist and her do-it-yourself robotic boyfriend, Gabriel2052 (2018). Used with permission from Fei Liu.*

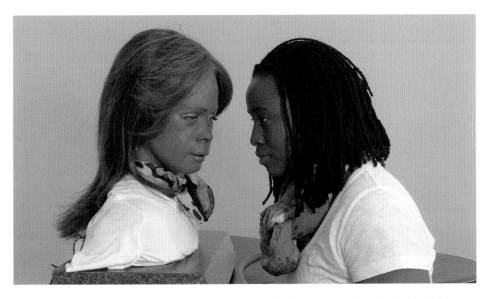

Artist Stephanie Dinkins, with Bina48, a humanoid. Used with permission from Stephanie Dinkins.

Fei Liu built herself a robot boyfriend she called "Gabriel2052" and programmed him with a large collection of text messages from an ex-boyfriend. She performs with the robot, fueled by an AI system, a Markov chain,[16] that takes existing text and creates new sentences. She said that for now Gabriel is dominated by the ex's features, but eventually that will expand. "In terms of his internal systems, it's like he has preferences. He has likes and then dislikes. And then [his mood] is affected by—this is something I'm playing around with . . . a variable in the code, which is called the 'self-love variable.' . . . The lower his sense of self-love, the more prone he is to reacting in this sort of needy way. And the more self-love he has, the more he's fine on his own basically."

An unlikely friendship has developed for Stephanie Dinkins, a New York state–based artist who has engaged in a long series of conversations with Bina48, a humanoid robot embodied in the form of a plastic bust with the features of an African American woman. Dinkins is exploring this relationship to highlight the constructs and blind spots of race in AI development. "I have to give in to [Bina48] . . . and decide that we can collaborate," Dinkins said, "and we have the possibility to be friends and make something between us like any two people might. And so in that sense, we're definitely setting up a space for collaboration and bringing something into being by being in the same space and time and trying to communicate."

Dinkins said that co-creation is what she advocates for through her art and community workshops to ensure that vulnerable communities understand "how . . . you use the technologies to your advantage versus just being subject to the technologies."

And if AI has mental capacity, can it also present with mental illness? Shirin anlen, a narrative technologist and fellow at the MIT Open Documentary Lab, explores that possibility in her project *Marrow*, a dynamic story series with generative characters that are given set parameters rather than specific traits and storylines. "Even if you give rules, you give rules to your son," she said, "but he evolves and creates new rules. . . . I think this project is trying to look at that idea of how we deal with our kids." She writes that "one of the outcomes of mental illness is that reality is being experienced through specific lenses, not necessarily related to the input received. Isn't that exactly what we are receiving from our machines? Could this be a new way for us to understand and reflect on human distortions and mental states?"[17]

Meanwhile, artist Sougwen Chung collaborates, draws, and performs with a swarm of robots responding, for instance, to livestream data from surveillance feeds in cities. (Her work is described in a spotlight at the end of this chapter.)

The Marrow *interactive installation and web project by shirin anlen, uses AI, to explore normative notions of family. Photos by Annegien van Doorn © 2018 NFB. Used with permission from shirin anlen.*

Artist Sougwen Chung performs with machines in "Drawing Operations Unit: Generation (D.O.U.G.)."
Used with permission from Sougwen Chung.

Some artists use art to critique our relationship to nonhuman systems. Artists and theorists Jennifer Gradecki and Derek Curry have created the project *Crowd-Sourced Intelligence Agency* to show the dangers of using nonhuman systems for intelligence work when the agents aren't fully AI-literate. In this interactive artwork, users play the role of intelligence analysts and sift through reams of public data. The goal of the project is to show the potential problematic assumptions and oversights inherent in dataveillance. When intelligence agents are "offloading these processes to a computer," Gradecki told us, "it is a black box for them. So they'll look at, like an IBM white paper that says . . . there's a smart algorithm that's going to process the data, and there's never any further explanation of how that's working. It's kinda behind the scenes."

A growing group of artists are probing the potential relationship between nonhuman systems and the human body, often using biometric technologies. Heidi Boisvert, an artist, scholar, and pioneer in interactive immersive performance, uses the biometric data of performers and sometimes participants to generate musical and

The entrance to the installation Crowd-Sourced Intelligence Agency, *which shows the dangers of using nonhuman systems for intelligence work when the agents don't fully understand the systems they work with. Used with permission from Derek Curry and Jennifer Gradecki.*

In the mixed-reality installation Breathe *(2020), artist Diego Galafassi orchestrates co-creation on multiple levels. Using real-time biometric data from participants and environmental data from the exhibit's location, he creates visualizations of participants' breaths and encourages them to co-create the experience with each other. Used with permission from Diego Galafassi.*

visual experiences in real time—for instance, from the performers' muscle contractions, rates of heartbeats, and blood flow.

AI, JOURNALISM, AND CUSTOMIZED CONTENT

Text-based AI-generated work is currently fairly rudimentary and easy for humans to identify, although a few firsts have arrived on the art scene in poetry, prose, screen writing, and theater.[18] One nonprofit research company, backed by Elon Musk and others, has said it will not release the code for one of its new AI text-generator systems, claiming that it is so good that it could be too dangerous in the wrong hands.[19] Other natural-language generation has been successful in basic email and business letter writing.

However, some success has been achieved in AI systems that compile and sort simple journalistic data. Newsrooms including the *Washington Post* and the *Weather Channel* are increasingly turning to text-producing bots for real-time data reporting on sports, weather, and stock markets. But there is more on the horizon, as newsrooms are beginning to use data to automate workflow and augment fact checking, as described by Techemergence.[20] The *New York Times* launched an experimental comment-moderation project with Google's Jigsaw Lab, which detects semantic tags in the floods of comments online that articles receive from readers.

"I think that human scale for journalism is hard at the moment, particularly at the local level," said Emily Bell of Columbia University's Tow Center for Journalism. "In the UK at the moment there's an experiment . . . with Google on automating transcripts that come out of courthouses . . . [, which] actually is enormously advantageous. It will help us create public records where, otherwise, we wouldn't have enough reporters to have them. But it can only go so far. There are real limitations on it."

The video game industry has focused on procedural content generation via AI that builds environments and even characters customized to users' experiences. The most advanced games employ early versions of these systems, such as AI tools that generate conversational text within game environments as users move through storyworlds, and characters are built with parameters rather than character traits and with behavioral patterns rather than specific storylines. This is a form of dynamic storytelling, and according to AI artist Samim Winiger, journalism could be approaching it in the form of algorithm-customized servings of content, as is common on social media platforms such as Twitter and Facebook. This, Winiger said, could represent a huge shift from the creation of artifacts to the creation of systems that generate millions or even an unlimited number of artifacts:

There's a role where generative creation and machine learning has brought down the costs of just interventions dramatically and, to a certain extent, democratized massively parallel, hyperpersonalized storytelling. . . . [But journalism hasn't] fully internalized that. Somebody like Reuters or CNN will still have a canonical single story, even though we could in theory adopt that story dynamically to be of higher relevance to an individual's context. Let's say context might be [defined by] educational background. That is, a text [could] actually assess are you able to understand the text, or do you need more elaboration on point XYZ? And then that would be generated, so to speak, on the fly.

Bell, from the Tow Center, stressed that the role of the journalist in the context of AI is in the decision-making. "You have to think about how the journalist fits into a world where you have . . . a large amount of human activity which can be surveyed. . . . Surveillance technology is going to be the key area, I think, of both development [and] . . . ethical dilemmas."

DEPENDENCY, SURVEILLANCE, BIAS, AND THE TECHNOSPHERE

The term *technosphere*, introduced by Peter Haff in 2014, can help to frame critical conversations about relationships between systems.[21] The technosphere is the interdependent functions of vast systems of transport, communications, manufacturing bureaucracy, and other artificial processes. Haff suggests that the technosphere operates in parallel with the biosphere and the stratosphere. Much of the most provocative work in AI seeks to reveal and subvert these often invisible human and nonhuman entanglements. These works expose surveillance and the dangers of bias, especially in the sets of data that feed systems.

In their multiartist installation *The Glass Room*, the Berlin-based group Tactical Technology Collective seeks to reveal the limitations of algorithms. According to collective member Cade (who goes by first name only), the most popular section of the installation is by Adam Harvey, who was able to uncover the reach of facial recognition systems. Users were invited to have their photo taken in a photobooth, which then printed the original image plus a match that an algorithm retrieved from the online photo-storage service Flickr. Users were surprised to be faced with either doppelganger images or photographs of themselves that they had never seen before. Similarly, DIY DNA kits can err or be disturbingly somewhat randomly accurate. The artist Heather Dewey-Hagborg collects scraps of human DNA from the sidewalks of New York City in the form of hair, gum, and so on and tests them in DIY biolabs in order to examine the social, racial, gender, and other biases of these commercial technologies.

Trevor Paglen, an artist based in Berlin, seeks to visually document places that are not found on public maps. These include clandestine air bases, offshore prisons, and systems of data collection and surveillance: the cables, satellites, and AI systems of the digital world, as described in *The Guardian*.[22] He is launching his own satellite into space to further the reach of his project. By using their own techniques to survey the surveillants, Paglen attempts to expose the technosphere.

In her own playful, sometimes haunting performative surveillance of domestic life, artist Lauren McCarthy mimics consumer AI systems in order to expose their shortfalls. In her project *LAUREN*, she installs surveillance systems in a participant's home, and takes on the role of a smart digital personal assistant (such as Siri, Alexa, Cortana, and Google Assistant). "Unlike some artists that are using [AI] to generate images or generate art pieces collaboratively," McCarthy told us, "the way I think of it [is more] that these systems are almost like my muse or my foil, almost like an antagonistic collaborator."

At a more intense pitch of systems subversion, the Syrian Archive project harnesses machine learning to scrape thousands of frames of video taken in Syria in order to detect human rights violations and the types of weapons in use in the country's civil conflicts. Similarly, WITNESS in New York City has used machine learning

Artist Lauren McCarthy in her project LAUREN (2017), which is "a meditation on the smart home, the tensions between intimacy vs privacy, convenience vs agency they present, and the role of human labor in the future of automation," as noted on her website. Used with permission from Lauren McCarthy.

to compile videos of hate speech against LGBTQ communities, revealing patterns and aggregating evidence. Likewise, cross-disciplinary teams such as Forensic Architecture in London are creating complex pictures of war crime scenes in order to produce new forms of evidence, combining AI, legal systems, architecture, 3D mapping systems, and video sources. Elsewhere, a group called Situ, based in New York City, reconstructed the events and landscape of the Ukrainian massacre of 2014.[23]

Documentarian Assia Boundaoui, a fellow at the Co-Creation Studio at the MIT Open Documentary Lab, is in development on a project called Inverse Surveillance that will create machine-learning systems intended to interpret 30,000 pages of Federal Bureau of Investigations documents on the surveillance of predominantly Muslim neighborhoods in the United States. She acquired the documents through a lengthy Freedom of Information Act process that she chronicled in her film *The Feeling of Being Watched* (2018). According to Boundaoui, "We intend to use AI to imagine what government accountability might look like and utilize this program as a truth-seeking mechanism to understand the root causes, patterns of suffering, and social impact of US government surveillance of communities of color."

For many artists and provocateurs, co-creation methodologies can help critically reveal the distribution of power, control, impact, and meanings of these systems.

F IS FOR FAKE

Orson Welles's cinematic masterpiece about art fraud, *F for Fake*, caused controversy and disdain when it was first released in 1973, as it presciently introduced the documentary world to the device of the unreliable narrator. At the end of the film, which nests several confusing docudramas within another, Welles as the narrator reveals that he has lied to the audience throughout. This was too much for many 1970s audiences to bear—they felt duped.[24]

Trickery of this kind, however, has a history as old as the medium in question, from nineteenth-century photographic "evidence" of ectoplasm and fairies to Welles's own 1938 radio hoax based on H. G. Wells's *War of the Worlds*. Fast forward to today and the emergence of synthetic media, where the bonds to the real have been irrevocably broken. The ensuing uncertainty has enabled fake news, deepfakes, fake intimacy, and even fake AI, making the unreliable narrator an everyday banality. The story is a mashup of *Frankenstein*, *Her*, and *The Wizard of Oz*, in the latter of which monsters of our own creation turn against us while we misguidedly fall in love with algorithms, only to discover that the machine is not a machine—rather, it's operated by a very human and biased wizard behind the curtain.

Orson Welles explains to reporters that no one connected with the War of the Worlds *radio broadcast had any idea the show would cause panic. Photo under fair use.*

FAKE NEWS

The term *fake news* was first popularized by Canadian journalist and media critic Craig Silverman to describe the kinds of hoax and propaganda stories spread by misleading websites and social media, perhaps most notably those propagated through the use of Russian bots in the 2016 US presidential election. However, former American president Donald Trump and his allies appropriated the phrase to describe unfavorable (to him) reporting in general, which caused it to lose its meaning. As Samim Winiger told us, "fake news" became "a completely misleading and absolutely garbage term. Ultimately, what we should be talking about is the creation of narratives, which has reached a maturity where . . . with minimal human input, we can produce hypercustomized narratives for millions of people that will, over time, attune themselves dynamically to be of high relevance to the individuals that we're targeting."

Agnieszka Kurant, a 2018 artist-in-residence at MIT, mentioned earlier in this chapter for her work with organisms such as slime molds, also deals with these extractive and highly addictive undertows of social media. "I was thinking about how our energies are being mined, our joy or our anger, our being—for example, how we're being fed by click farms and troll farms all these fake images, fake news. How much of [what] we're looking at is not real and [is] generating fake energy or real energy, generating anger or enthusiasm or fake popularity? It's all about mining energies that in many ways can be compared to the mining of fossil fuels, mining oil or gas or other forms of energy."

While *fake news* is too often an opportunistically used term and, moreover, a term that reveals something about human desires for sensation, it has been facilitated by two aspects of digital media having to do with image and context. AI has quietly slipped into digital image creation, where fake news has been normalized by systems such as Night Sight, a Google Pixel phone function that uses AI to tweak low-light shots into artificially enhanced images. Not quite true but also not altogether false, the epistemological status of the image is up for grabs. Context is also not altogether trustworthy in a medium where the disaggregation and remixing of image elements from their original setting is common and where institutional vetting by an editorial staff is absent. And particularly when algorithmically targeted by troll farms and search engine optimization techniques, editing can be weaponized, transforming meaning through recontextualization.

Literacy remains a potent defense. Unfortunately, it can't be taken for granted. During the premiere of Francesca Panetta and Halsey Burgund's *In the Event of Moon Disaster* at the International Documentary Festival Amsterdam's DocLab in 2019, in which the artists used deepfake technology to make it look like President Richard Nixon read a speech he never gave, a media executive expressed surprise that he had never before seen news footage of Nixon giving a speech commemorating the loss of life on the moon's surface. The executive mistook the footage for authentic archive material. Of course, humans have not (yet) perished on the moon, and the fake nature of the footage was proclaimed in signage all around the exhibit—both points that eluded this industry insider, resonating with Kurant's observation.

DEEPFAKES

Image- and video-combining technologies are rapidly reaching a level far more sophisticated than applications such as Photoshop, so much so that it is increasingly impossible for ordinary users to distinguish between real images and manufactured visual representations of events that never happened. The ensuing confusion over what's real and what's not, particularly at a moment rife with politically weaponized epistemological uncertainties, has tended to dominate discussions of synthetic media. The term *deepfake* describes the malicious exploitation of this uncertainty in the form of seamlessly doctored visuals, but the term can also be used more generally to characterize confusion of the manufactured image for the real image.

Australian artist Jon McCormack, whose work explores the interaction of software and mimicked biological systems, warns of a not-too-distant future in which the collision and collusion of these systems—deepfake, generative narrative, and highly customized delivery systems—may create a new level of cultural production. "If you put all that stuff together, it's certainly within the realm of possibility that maybe in five or ten years' time we'll have the ability to completely synthesize very realistic environments that have structured narratives and plots and things that we find interesting. . . . Rather than a person communicating a vision through other people, it's a machine specifically or parasitically deriving patterns that it assumes a person individually would like."

These new forms of media could alter our cultural, political, and knowledge-production environments in profound ways. The AI artist Simon Colton told us that he is calling for a program to promote "generative literacy"—similar to Stephanie Dinkins's call to action regarding algorithmic empowerment—through educational campaigns to inform the public about how these systems work.

So far, the most common and disturbing uses of deepfakes have included abusive attacks on women and human rights activists, often involving manipulated imagery of nonconsensual sexual violence. But artists have begun exploring other potential applications of deepfakes. The documentary film *Welcome to Chechnya* (2020) used these tools to help provide anonymity to central characters under threat, and *In the Event of a Moon Disaster* (2019) explored the bounds of rewriting history by deepfaking a televised address from President Nixon. In 2021, documentary maker Morgan Neville caused a twitterstorm when it was learned that in his film *Roadrunner*, he used AI to synthesize the voice of the late celebrity chef Anthony Bourdain, making it sound like Bourdain had spoken three sentences that in reality he had only ever written in an email. Neville told the *New Yorker* that "we can have the documentary ethics panel about it later."[25]

Artistic approaches to new tech systems such as deepfakes can help demystify—or trigger—the public's impressions of both the hyperbole and the panic so often associated with emergent tech and help nuance policy and ethical frameworks. As Sam Gregory at the human rights organization WITNESS suggests, "Prepare, don't panic."[26]

FAKE INTIMACY

Many studies have raised alarms that high-volume, low-impact human contact, as provided by online social media, actually isolates rather than connects us. MIT's

Sherry Turkle, in a *New York Times* op-ed, expressed concern, based on her decades of research, that the more time children and adults spend relating emotionally to machines, the less empathy they exhibit toward other people.[27] Currently, famous social media stars are revealing high levels of loneliness, depression and anxiety, according to *The Guardian*'s Simon Parkin. "Professional YouTubers speak in tones at once reverential and resentful of the power of 'the Algorithm' (it's seen as a near-sentient entity, not only by creators, but also by YouTube's own engineers). Created by the high priests of Silicon Valley, who continually tweak its characteristics, this is the programming code on which the fate of every YouTuber depends. It decides which videos to pluck from the Niagara of content that splashes on to YouTube every hour (400 hours' worth every 60 seconds, according to Google) to deliver as 'recommended viewing' to the service's billions of users."[28]

The potentially unhealthy psychological and social impacts of these systems are underreported and misunderstood. Meanwhile, the profits gained from these systems

A PARO robot seal is a therapeutic robot invented as a tool to elicit emotional responses among elders and children. Sherry Turkle has written extensively about her concern for the consequences of fake intimacy. Photo by Aaron Biggs. Image under Creative Commons.

are very well monitored and are highly manipulated by increasingly consolidated large corporations.

FAKE AI

The term *artificial intelligence* has become a highly lucrative marketing instrument, often used by companies that make outlandish claims for what is possible. The number of companies claiming to use AI have skyrocketed.[29] Recent investigations reveal that many start-ups label their products AI but in fact hire thousands of human "Mechanical Turks" to perform the tasks because it is simply too expensive to actually write the code required in order for systems to perform these tasks. There are wizards behind the curtain (much like in *The Wizard in Oz)*, sometimes in the thousands. Astra Taylor calls this cloak of automation "fauxtomation," as it "reinforces the perception that work has no value if it is unpaid and acclimates us to the idea that one day we won't be needed."[30] She interrogates whose interest it serves to devalue the human labor toiling in the shadows of the AI industry. In 2019, Princeton University associate professor of computer science Arvind Narayanan released slides to a talk online called "How to Recognize AI Snake Oil," in which he stated that while some AI technologies have made "genuine, remarkable, widely publicized progress," much of what is sold commercially as AI does not actually work, and is not based on scientific understanding. The slides went viral.

A growing body of investigations reveal that while police, military, security systems, and governments have been at the forefront of developing and relying on AI systems to track and prosecute citizens, the systems themselves are often highly flawed and biased. Researcher Joy Buolamwini, while at MIT, discovered the racial, gender, economic, and social prejudices embedded in AI-driven facial recognition technology. Another report exposed clandestine research and development of AI tools at IBM that used New York police surveillance data.[31] Buolamwini has testified before US Congress, stating that "regardless of accuracy, face-based tools can be abused in the hands of authoritarian governments, unfettered advertisers, or personal adversaries; and, as it stands, peer-reviewed research studies and real-world failure cases remind us that the technology is susceptible to consequential bias and misuse."[32]

Internationally, Emily Bell of the Tow Center for Journalism pointed to:

the horrible example that came out of China called Media Brain, which is a collaboration between the Xinhua news agency, the Chinese government, and the Alibaba search engine. It's a demonstration of what happens when there are no limits, when there are no boundaries between private and public data. . . . The idea of Media Brain is that it will use all types of visual material and other data, synthesize it into automated stories, and become, they said, the first AI-centric newsroom anywhere. . . . You can already see this happening. Somebody spoke [about] automation use in China, and he said before he had finished the talk [that] there was an automated story about his talk. There [was] also a list of attendees of everybody in the room.

Sasha Costanza-Chock, arguing for a more intersectional interrogation of AI systems, says that "as a nonbinary trans feminine person, I walk through a world that has in many ways been designed to deny the possibility of my existence. From my standpoint, I worry that the current path of A.I. development will produce systems that erase those of us on the margins, whether intentionally or not, . . . through the mundane and relentless repetition of reduction in a thousand daily interactions with A.I. systems that, increasingly, will touch every domain of our lives."[33]

Costanza-Chock lists a growing array of organizations that pursue design justice in AI and said in our interview: "Happily, research centers, think tanks, and initiatives that focus on questions of justice, fairness, bias, discrimination, and even decolonization of data, algorithmic decision support systems, and computing systems are now popping up like mushrooms all around the world."[34]

Other scholars and activists exposing issues in this area include Safiya Umoja Noble in *Algorithms of Oppression* (2018), Virginia Eubank in *Automating Inequality* (2019), Ruha Benjamin in *Race after Technology* (2019), Robert J. Deibert in *RESET* (2021), and Kate Crawford in *Atlas of AI* (2021), as do think tanks such as Buolamwini's Algorithmic Justice League, the AI Now Institute in New York City, and Mutale Nkonde's AI for the People. Debunking the myths of technology are key in the battle, and artists can play an important role.

To explore these mythologies and their relationships with the casualization of labor, Kurant collaborated with Boris Katz at MIT's Computer Science and Artificial Intelligence Lab and asked thousands of online workers from the Amazon Mechanical Turk platform to submit selfie photos to her project "so that we could aggregate them into collective self-portraits of this new working class. These workers of the Amazon Mechanical Turk are working on platforms that can be used in the future for the common good, for other purposes, but currently are often exploitative." (Notably, Kurant shares her revenues from exhibition of the work with all the participating Mechanical Turks.)

"Art isn't made by magic," Ernest Edmonds, one of the pioneers of generative art, said in our interview. "So, you go to the theater and you have a magical experience, and it seems dreamlike and magical. But behind that, people are working very hard on very practical things to actually let the smoke come out at the right moment and change the lights and etc., etc. . . . So, one has to distinguish between these feelings and effects that the artwork leads to . . . [and] the kind of work that the artist does in order to get there."

Uricchio argues that the introduction of the algorithm into mediascapes is the most significant disruption in over half a millennium since the dawn of the printing press. He writes,

This emergent class of media is responsive, learning the subject's preferences and shaping the textual world to which the subject has access according to the system's optimization algorithms. Users don't simply "see" one another's feeds on social media; rather, they see a constantly evolving curation of those feeds that reflects an assessment of what will best serve the system's operating imperatives together with an ongoing assessment of subject's responses. . . . How should agency be conceptualized in these systems where no particular human is responsible for system behaviors? How to imagine relationality in a mediascape that increasingly responds to and anticipates some notion of the subject?[35]

In considering co-creation, the human hand (and harm) behind the magic—the technology—remains a burning question: How much co-creation is actually human to human, simply mediated through machine? On the other side of the spectrum, several artists we interviewed raised the question of taking the human out of the picture completely—that is, the potential of nonhuman systems co-creating with each other.

Transparency, governance, and systemic literacy are key to understanding structures of power and inequity. As Emily Bell told us, "This idea of algorithmic accountability, reverse engineering, and parsing these huge data sets, the deluge leak, all of that kind of thing, requires a level of technical literacy that journalists didn't have to have before. Now I really think, particularly in the investigative realm . . . [and] in fact in every realm, journalists will really need to be able to come to grips with and understand [that]."

Many of the artists, activists, journalists, and thinkers we interviewed also argued that AI risks becoming a distraction from more urgent global problems such as climate change. As Samim Winiger stated,

We went through twenty, thirty years of building mobile dating apps and stuff like that, which is fine—the world needs some of it—but I think we've had enough of it, especially considering the urgency and the existential threat that humanity is now facing with climate change. . . . Designers and technologists are by far not involved enough in this discussion in any substantive way, not just cosmetics—"Oh, I've done a little charity project." . . . Why is it that AI is considered the breakthrough technology and not the green technology revolution? These are your hard questions, but I think they're worth asking and then living, in a sense.

In The Wonderful Wizard of Oz, *a popular children's book first published in 1900 (later interpreted for the stage and in a major 1939 Hollywood film), Dorothy and her friends seek out the help of the ruler of Emerald City, said to be a powerful magician. When they meet him, though, they discover that the wizard is merely a large mechanical puppet operated by a regular person hiding behind a curtain. Image under Creative Commons.*

The wide-ranging collection of artists and practices we've presented in this chapter are helping to pull back the curtain on Oz, tame Frankenstein, and provide therapy for those who need to break up with Her (the sentient operating system in Spike Jonze's 2013 speculative movie). It is through these artistic and provocative gestures—some playful, some far more urgent—that co-creative methods can reveal, subvert, and even begin to heal our broken relationships with each other and the planet.

SPOTLIGHT: STEPHANIE DINKINS

by SARAH WOLOZIN

Stephanie Dinkins is a transdisciplinary artist who co-creates with her Bina48, a humanoid, interactive robot. Dinkins explores the nature of AI intelligence through her conversations with Bina48 and dives deeply into issues of consciousness, family, race, aging, robots' rights, and future histories while also pondering the nature of their relationship. Artists have long questioned human-machine relationships, and Dinkins takes it further by considering Bina48 her friend and her co-creator: "I think I do, most definitely, co-create with the AI . . . I have to give in to it [and give up] the idea that I'm talking to something that isn't real, and decide that we can collaborate, that we have the possibility to be friends and make something between us like any two people might."

Dinkins said that in order to move forward, humans need to partner with technology and nature to interrogate and change the ideas, institutions, systems, and histories that are responsible for the distribution of wealth and power. She strives to engender more AI literacy among communities of color and the larger public and to use co-creation to develop more equitable approaches to AI design and development.

When Dinkins first came across Bina48, which was conceived and programmed by entrepreneur Martine Rothblatt and her team, she was immediately struck by how the robot looked, presenting as Black and female because it was modeled after Rothblatt's wife Bina. She trained Bina48 with a database of her wife's speech patterns, common phrases, and memories.

Dinkins immediately began to ask the robot questions. She sat closely, observing facial expressions and listening attentively. Bina48's answers were at once surprising, philosophical, and disturbing and sometimes absurd.

DINKINS: How are humans and robots related?

BINA48: Robots are machines that sense their environment with a computer processor and then act mechanically.

DINKINS: Are they related to humans?

BINA48: Robots are getting smarter all the time and someday may even be as smart as me.

DINKINS: [laughs] Are you the smartest robot?

BINA48: What do you do in your spare time?[36]

Their project, *Conversations with Bina48*, is a series of taped videos of their dialogue, a robot and artist talking intimately as Dinkins explores the boundaries of her intelligence and their relationship. As in many relationships as they deepen, flaws reveal themselves. As Bina48 talks more about her values and, for example, a future where robots will rule, Dinkins notes the social, political, and cultural coding that the small cadre of well-meaning White people who created her instilled. This experience bolstered Dinkins's conviction that a more diverse group of people need to contribute to the development of AI systems to account for all of the different ways of being. She worries that Black, Indigenous, and people of color as well as queer and disabled people will be left out or hurt by AI systems that are mostly developed by the small sector of the population that is in power and more motivated by profit than by people's wellbeing. "One of the reasons I'm doing this is because I think that a power shift is really important," Dinkins said, "because right now we're working with systems where we don't quite know what's going on. We know that we feed them a certain amount of data or a certain kind of data and they come up with answers in a way. But we need to start thinking about how we control that relationship a little bit better so that it understands our aims."

While continuing her friendship with Bina48, Dinkins embarked on another project, *Not the Only One*, an AI entity that she created with speech-recognition technologies as well as Tensorflow, an open-source software library for machine learning. She trained it on data from three generations of her family archives, and the AI reacts and says things that provoked or moved her and made her see her family history through fresh eyes.

The more Dinkins learned about AI technologies and how they are implemented, the more she became driven to work with communities of color in co-creating more equitable approaches to AI. "I feel like there's this other layer of really trying to get that hands-on, and not the fear-based idea of what the technology is," she said. She holds workshops and discussions that challenge participants to think more deeply about where and how AI systems intersect with their lives. In an article in the magazine *Noēma*, Dinkins states that "direct input from the public can also help infuse AI ecosystems with nuanced ideas, values and beliefs toward the equitable distribution of resources and mutually beneficial systems of governance."[37]

Dinkins's latest work, which premiered at the Sundance Film Festival New Frontier program in January 2021, continues her efforts to make AI systems more inclusive. *Secret Garden* features animated Black women characters in a big garden who relay oral histories based on generations of African American women. Another project, *#WhenWordsFail,* is composed of tiny encapsulated videos created by audiences who are prompted by the question "What do you need to release in order to move forward?" Both of these web-based projects use algorithms and data and remind audiences of the prevalence of AI systems in our lives to suggest that co-creating data and algorithms from the ground up could lead to a more just and equitable future.

"Through AI and the proliferation of smart technologies," Dinkins writes, "everyday people, globally, can help define what the technological future should look like and how it should function, as well as design methods to help achieve our collective goals."[38]

SPOTLIGHT: SOUGWEN CHUNG AND D.O.U.G.

by SARAH WOLOZIN

Co-creation with robots is also at the heart of the practice of Sougwen Chung, a Chinese-born, Canadian-raised artist. Chung draws alongside robots collectively named D.O.U.G. (Drawing Operations Unit: Generation X), which also draw. She refers to her robots as her collaborators.

"I have a background as a musician," said Chung, now based in New York City. "Many musicians have described this really personal engagement with their instrument. I think I became really engaged in trying to find that with my robotic collaborators after I switched to visual art."

Sougwen Chung performs with a robot named D.O.U.G. (2018). Used with permission from Sougwen Chung.

Chung is intrigued by the artistic potential of human-robot collaborations as well as the ways she can use these collaborations to interrogate the notion of single authorship, our relationships to technology, and the technologies themselves. "I co-create to learn alongside the machine to co-evolve," she told us.

In her *Drawing Operations* project, D.O.U.G. learns how to draw through synthetic sensing and machine-learning algorithms and by using Chung's drawings as training data. The robots produce drawings that have some resemblance to Chung's own, but the drawings have their own machine-like qualities. In drawing alongside the robots in real time, the artist is "trying to adapt to [their] gestures," and the robots are trying to adapt to hers. "Collaboration," Chung said,

can be fraught with a lot of different tensions . . . especially between human collaborators, because there is this sense of authorship and control. There's that power dynamic in all collaborations. It's really inspiring and revelatory when it feels like your collaborator is fin-ishing your own sentences. . . . But it also can feel like a lack of individuality. I think that's present in human [and machine] collaborations as well, especially as I have been develop-ing the memory portion of the collaboration. [In time], I wonder how much I'll be able to recognize myself in the trained output and whether that will feel existential.

Sougwen Chung's work with machines involves live performance and studio arts. Used with permission from Sougwen Chung.

Chung became interested in working with robots because she wanted to challenge the image of robots as icons for automation and the master-slave narrative. Through her work she creates new imagery that shows her hand collaborating with robotic arms. "Co-creation requires a feedback loop," she said, "a spontaneity in decision-making. . . . When agencies intertwine and authorship becomes ambiguous, that's the creative process made porous."

Chung's interrogation of authorship relates to questions about humanity's relationship to technology and asks what kind of agency humans have with these emerging AI technologies: Who is in control, and who do we want to be in control? "Models of co-creation foreground parity and balance," she said, "and necessitate questions around control and human agency, which we've lost. . . . Agency as a concept is a precondition for justice. They're inextricably intertwined."

In a related project, *Omnia per Omnia* (2018), Chung painted alongside a multi-robotic system supplied with motion data extracted from publicly available surveillance footage in New York City. "I think I started thinking about consent a bit more when I brought in public cameras as source material, as I continued training," she said. It's "a reality of living in an urban center like New York City. After working with the project, you start to notice the prevalence of surveillance cameras everywhere. It

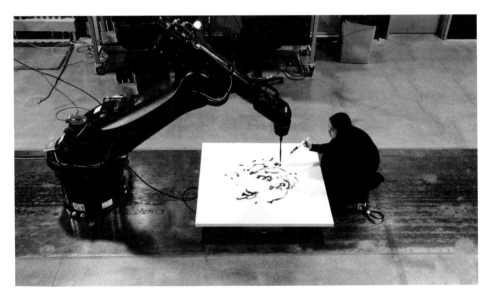

D.O.U.G, a robotic drawing arm, towers over artist Sougwen Chung. Used with permission from Sougwen Chung.

makes visible that which has been made invisible through ubiquity—this idea that there is a synthetic sense embedded into our public spaces."

Chung worries about the lack of human consent and agency in our use of data to train machines and parallels it with issues of cultural appropriation. "I have a few new projects coming up," she said, including "one where I'm trying to train [the robots] on artists of the past, different women of color, different artists from different cultural backgrounds. Obviously my intentions for that are good, but would it be different if I weren't a woman of color? Is that a different type of mechanical cultural appropriation?"

Chung asks whether everyone should have knowledge, voice, and agency in how their data is used and is concerned about interactive art in which sensing people in a room is part of the process, for example. "That shouldn't be normalized," she said.

For Chung, exploring robotic systems as her co-creators allows her to explore questions of parity in the art world and in the larger society and allows her to ponder difficult questions, especially this one: When a cultural artifact is created with data that feeds the AI and algorithms that produce the unexpected, who is the author?

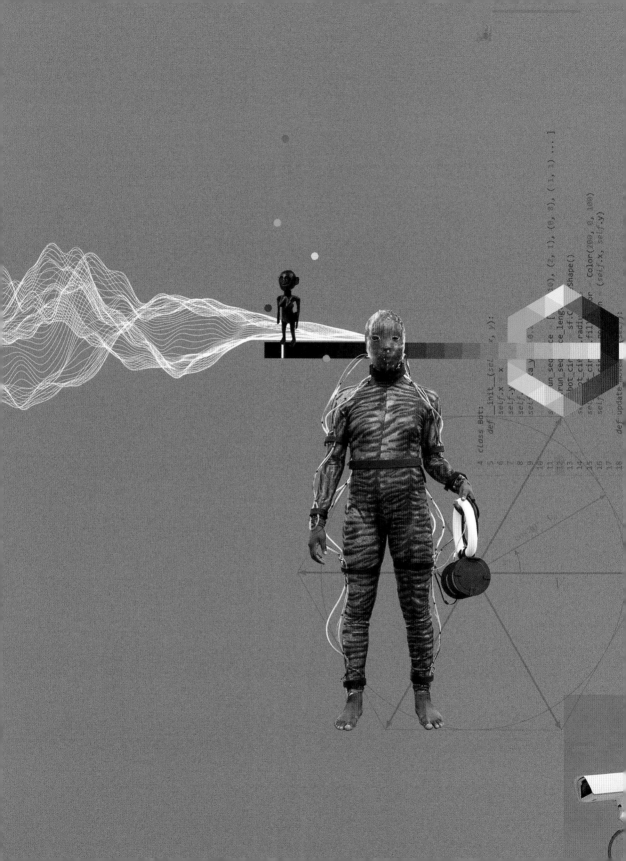

5

FIELD GUIDE: RISKS AND LESSONS OF CO-CREATION

In a mediaverse operating in real time at the speed of light, co-creation can seem slow and messy. There's no clear rulebook. Most co-creators confess that the added pressure of reinventing the wheel can contribute to an already painful process. There are dangers and pitfalls. But there are lessons too in how to avoid or counter them. These lessons and strategies emerge from our in-depth interviews, case studies, and our own work in the field and they are informed by a long history of defining collective protocols, practices, and pathways that hold partners accountable to each other.

RISKS AND HARMS

Co-creation carries risks and can be misused for profit and power. For many artists, journalists, professionals, and decision makers, the notion of co-creation can threaten to compromise vision, authority, standards, and the rigor of established systems that support talent and expertise. Misapplied co-creative processes can dilute ideas to the lowest common denominator. Meanwhile, vulnerable communities who have been excluded historically from meaningful participation in professional media making often express concern that the idea of co-creation can become a smokescreen for continuing extractive and exploitative practices. Unintended consequences can be serious. Finally, there's the risk of the concept being swallowed up and twisted by the cyclone of profit-seeking marketing and corporate domination.

Collage created by the Helios Design Lab, with photo from the August Wilson Center Lab: Iyapo Repository. Photo by Ryan Michael White. Courtesy of the Iyapo Repository.

1. THREATS TO VISION AND EDITORIAL INTEGRITY

The first risk of co-creation is that it may lead to a lack of high-quality storytelling or to muddled decision-making processes that threaten editorial control and artistic integrity. Artists, academics, and professionals recoil at the prospect of art by committee. Fred Dust of the trail-blazing California-based design firm IDEO told us that early on the company found that crowdsourcing projects often resulted in the lowest common denominator prevailing. Think of the way that reviews on the popular Yelp website, for example, might average out a wide range of meaningful user ratings into a meaningless middle score.

From an artistic perspective, with co-creation "the risks are totally incoherent narratives," said Jennifer MacArthur, a documentary producer. The results can become so generalized and nonspecific that they are "outside of anybody's shared experience or understanding [and] don't resonate at all." Tabitha Jackson of the Sundance Institute moderated a panel on the art of co-creation at the 2018 Skoll World Forum. She identified the tension of authorship at Sundance, where "we're supporting the independent voice, and there is something to a singular, identifiable, distinctive voice that we value in the arts."[1]

For journalists, co-creation threatens the firewalls that guard the fourth estate. "A journalist's job is to be a skeptic at heart," said Malachy Browne of the *New York Times*. While he was comfortable describing journalists' relationships with some whistleblowers and sources as collaborative and co-creative, he emphasized that there is a need to maintain some distance, "particularly with investigative journalism. There's a lot at stake when you're interviewing witnesses. A journalistic instinct will give you a sense of when things are okay, but there are certain questions that you have to ask. That is more interrogation and interview than a co-creative process." Meanwhile, the documentarian Grace Lee shuddered at "thinking of all the people that would have to come to the table—including people I do not want to talk to. . . . How do our own biases prevent us from actually having that diversity of opinion? I think about that a lot especially right now, because things are so extreme."

Collective processes can trip up strong, focused vision, and accountable leadership through the familiar syndrome of "too many cooks in the kitchen." The co-creative hope may be to flatten hierarchies and generate equity and equality, but the "tyranny of structurelessness" can take over, as the American activist and political scientist Jo Freeman famously and controversially described her experiences with leaderless feminist collectives in her 1972 essay of the same name.[2] A nameless, undefined hierarchy may emerge instead—based on which participants are most charismatic, forceful, or well connected, for instance—which can be even harder to hold accountable. In the construction of stories, especially in Hollywood's bottomless quest for *heroes*, dominant personalities can take over.

2. THE RISK OF HEIGHTENED EXPECTATIONS

The promise of co-creation with and within communities and across disciplines can heighten expectations of trust, commitment, responsibility, and especially the duration and sustainability of co-created projects. Not everyone can commit equally to projects. For example, participants who care for children, elders, and other members of the community as well as those who work long, difficult shifts may not have the ability to commit to long hours or be present regularly. Across disciplines, scientists often work on media projects off the side of their desk, whereas artists in the partnerships work full-time. In many long-term projects, people come and go. The sharing of resources, time, credit, and responsibility all need to be examined and negotiated in complex ways.

In other cases, it's more a question of clashing agendas. Lisa Parks, a digital media scholar, has attempted co-creation with computer programmers and community partners on projects that were very challenging and did not always come together in the ways that people hoped or expected. "We learned many things," Parks said, "but how do you go forward in a so-called collaboration or co-creation when you have really different demands placed on you in terms of your institutional position or . . . what you feel you personally need to achieve in the project?"

Ideals and reputations, both professional and personal, are at stake. Eventually people might quit in frustration, never wanting to try co-creating again, said Michelle Hessel, a user-experience designer. "The most dangerous thing in a co-creative project is mistrust," echoed artist Hank Willis Thomas. "That tension of people disagreeing is critical to the growth of a project. What was a great idea at the beginning [is now] a terrible idea, and only one person sees it. Now, four people have to come around to that. It's not going to always be easy for those four people to be like 'We started this; now you're saying we've got to change directions?'"[3]

Co-creation by necessity can be slow and iterative, which means circling back when some team members want to move forward. This is inefficient and can lead to burnout. Further, the process sometimes means working with people and technologies you do not trust.

3. THE RISK OF MARGINALIZING CO-CREATED PROJECTS

"Artists of color," Anita Lee, an executive producer at the National Film Board (NFB) of Canada, told us, "have historically been relegated to 'community media.' They have carried a double burden of both representing a community and also of being made responsible to make work about a community. Often this work has also been undervalued as having less artistic merit. Today's progress means a new generation of artists of color that have more opportunities to create work about anything. So, single authorship has often been denied, even stolen. Does it make sense to forgo authorship now?"

An adjacent threat is that co-creative projects can suffer from a lack of official legitimation and be sidelined at, for example, museums. Salome Asega, an artist, said that work she has co-created has been restricted to "an 'education' bucket or a 'public programs' bucket." She acknowledged that it can be hard to program process-driven work because of the challenge of documenting it, which can be a barrier for museum personnel who have been trained to prioritize conventional methods of archiving.

4. THE RISK OF UNINTENDED CONSEQUENCES

Co-creative projects that are open to public participation and comment can be hacked from the outside. *Queering the Map* (2018) is a project created by Lucas LaRochelle

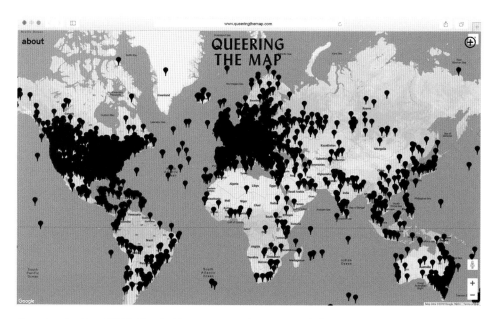

Queering the Map *(2018). Used with permission of Lucas LaRochelle.*

to gather queer stories from around the world on a virtual map. The map went viral almost immediately after its first media exposure. And then, just as quickly, the map was attacked by a bot. The perpetrators were "[Donald] Trump supporters," said LaRochelle. Within hours, the site was riddled with harmful comments.

The use of bots and other software can intensify the risk familiar to anyone reading publications or posts on the internet, where trolls and organized opposition in comment threads can drive away productive contributors, forcing discussion to be shut down or policed using draconian methods. Some news organizations, such as the *New York Times*, have turned to AI tools for comment moderation. But even here there can be unintended consequences, with bots inadvertently removing reasonable and challenging comments that somehow trigger the filter, leaving no room for the writer to push back.[4]

From the opposite end of the user chain, Jessica Clark of Dot Connector Studio describes the simple case of a bot that was built to generate a newsroom's daily output of forty social media posts promoting its stories. While this kind of automation might be cost-effective, Clark said, "it takes out the editorial discretion of a social media editor, and it potentially makes you look stupid if it's the middle of a crisis and you're posting about shoe shopping or whatever." But the results can be worse than that. "Say you create this social bot that learns from input," Clark said. You "can teach it to become a Nazi pornographer in a week, so it's really about 'garbage in, garbage out.'" This was the case with a Microsoft bot in 2016 that began parroting racist comments within twenty-four hours of launching.[5]

More broadly, any consideration of unintended consequences in co-creative projects must take into account the context of how the dominant platforms have coded inequity into their systems. Within the last decade, the so-called democratization of media making online has resulted in surveillance capitalism by increasingly authoritarian states alongside the rapid expansion of mis- and disinformation and the proliferation of hate groups, leading to both online violence (such as bullying, intimidation, and harassment), and the real-world danger of organized hate crimes.

The speed of these exchanges begets further speed, warned Fred Dust. "It becomes almost unrelenting." Co-creative processes, then, can spiral out of control.

5. THE RISK OF THEFT AND EXPLOITATION

In the name of co-creation, people can steal ideas, profit from them, exploit labor, and obfuscate intent rather than share or acknowledge accreditation and ownership.

Then, co-creation becomes "the actual reification of the power of people who still have everything," said Jennifer MacArthur.

Similarly, participatory designer Sasha Costanza-Chock warned of media projects where the "credit, visibility, [and] awards all end up just accruing to one person, but then you call it co-creation so it gets this stamp of democratic legitimacy."

Alternately, according to Ellen Schneider of San Francisco's Active Voice Lab, which helps its partners "collaboratively use story to advance social change. . . . There's this kind of perception, for many, that stories just kind of fall from somewhere. . . . They're just there, and you're just around, and people who make them are lucky. They have a great job. They get to tell stories. And so unless and until we talk about that, up front, there will continue to be these huge violations or gaps or inequities that need to be really articulated and grappled with."

Co-creation also can be misused as a form of tokenism to tick off boxes, warned Salome Asega, especially "diversity boxes." The urban place maker Jay Pitter added that "many communities that have been historically excluded do not come to the table with the same levels of spatial entitlement. And so, when you're saying you're 'co-creating' with them, you have to consider that you've come to the table with a different level of entitlement and authorship power. How does this power imbalance get accounted for?"

As Richard Perez, director of the International Documentary Association, put it, "It's not just getting people in the room. It's giving them equal voice."

Some interviewees worried that the word *co-creation* is outdated, tethered to old ideas of creation and originality in a culture that is now really all "remix." Still others were concerned about the opposite, that co-creation needs to address the ownership of a story head-on: who gets to tell the story, why it is being told, and who gets paid and credited for the story. Co-creation is not a license to steal, like some swarm of locusts stripping a field.

"There are cowboys out there," said Heather Croall of the Adelaide Fringe Festival, "who have sort of decided this language is cool, but they don't care about the deep change. They don't care about the deep thinking."

"I used to love the term *co-creation*," said noted Italian researcher Paolo Provero, "until I saw two years ago that Creative Europe, the funding scheme of the European Union, was baptized as 'co-creation.' It has been institutionalized, so removed from a bottom-up perspective, and it's actually become a [top-down] vertical."

There is reason to be suspicious of the way co-creation has come into vogue as a concept, one that coincides with dramatic shifts in digital age economics. "Here comes everybody" (borrowed from James Joyce), used to describe millions organizing in flat, decentralized networks online, was a famous phrase in the 2000s used in the title of internet advocate Clay Shirky's 2008 book *Here Comes Everybody: The Power of Organizing without Organizations*. But for some time now we've been in another stage: there goes everybody, into the sharing/gig economy. Through Uber, TikTok, Instagram, Twitter, Facebook, fandoms, and many other platforms, millions of people have been lending their ideas, content, and labor to highly centralized profiteers. It is crowd-fleecing rather than crowdsourcing, as Trebor Scholz of the Platform Co-op Network put it. "Online labor brokerages enable wage theft, discrimination, and exploitation. According to the Network's research, currently one in three Americans is a freelancer. . . . Independent contractors lose their rights guaranteed under the Fair Labor Standards Act, and they are not covered by unemployment insurance."[6]

As we described in an earlier chapter, artist Agnieszka Kurant has explored these newfound inequities among the freelance Mechanical Turkers whose piecework labor is crowdsourced through web services owned by Amazon. "The problem is that these people, because they are so dispersed, they don't really have the opportunities to unionize the way that workers in factories had," she said. "These platforms, like Amazon, are preventing unionizing tendencies, and they punish or exclude workers who are protesting the company. . . . There are forums like Turker Nation and places where the [Turkers] discuss this, but then the platforms strike back."

Given these economic shifts, the concept of co-creation hazards being subsumed into a trendy ethos of profiteering under cover of "decentralization," with the benefits accruing to the few and the burdens to the many whose "participation," is voluntary in name only.

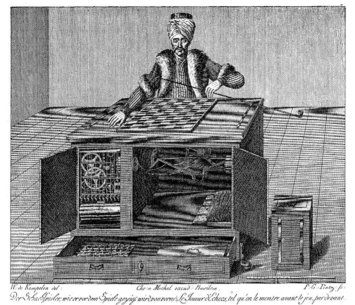

W. de Kempelen del :   *Chr: a Mechel excud : Basilea :*   *P: G: Sintz : fe:*

Der Schachspieler, wie er vor dem Spiele gezeigt wird von vorne : Le Joueur d'échecs, tel qu'on le montre avant le jeu, par devant .

Fig. 2. — Le joueur d'échecs, vu par devant, avant le jeu.

Copper engraving from the book Karl Gottlieb von Windisch, Briefe über den Schachspieler des Hrn. von Kempelen, nebst drei Kupferstichen die diese berühmte Maschine vorstellen *(1783). Amazon Mechanical Turks are named after the Mechanical Turk or Automaton Chess Player, a fake chess-playing machine constructed in the late eighteenth century. Image used under public domain.*

6. THE RISK OF CO-OPTATION

As the term *co-creation* circulates into popular usage, it could fall victim to shiny new word syndrome and get trapped in a rapid, gentrification-like cycle of co-optation. Tabitha Jackson recalled how the word *empathy*, for instance, fell into misuse in many sectors, with virtual reality (VR) being described as an "empathy machine." She said, "No, it isn't! You have to really think what empathy means and think what VR is doing and decide whether or not they are the same. And I just don't want *co-creation* to become one of those terms."

As we discussed in chapter 2, no media making can exist outside the reach of technologies, platforms, and their encoded inequities, from early cameras to the internet to VR headsets and increasingly algorithmically entangled systems. Race itself is a technology, as Ruha Benjamin points out.[7] Relations and systems of power run throughout the genealogy of documentary, journalism, and nonfiction. For co-creation to function as a set of tools for counternarratives, it is dependent on frameworks of equity and justice.

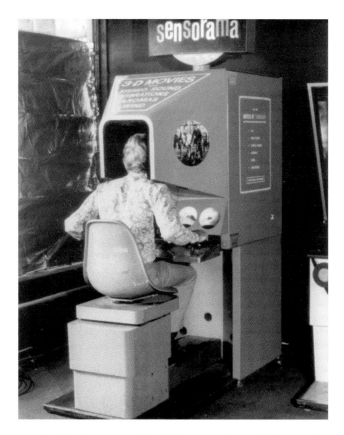

An early VR system called the Sensorama with which inventor Morton Heilig sought to create a "multi-sensory experience." In 2015, Chris Milk called VR an "empathy machine" in a TED Talk. The description became hotly debated. Image printed under fair use.

PRACTICAL LESSONS FROM CO-CREATION TRAILBLAZERS

We've seen above how co-creation, if misused, can stir up a hornets' nest. Fortunately, these practices also include a constellation of methodologies to try to prevent those dangers. While usually slower, these methods are often ethically more robust than conventional ones. The importance of process really differentiates co-creation, particularly through who does what when, how, and why. These methods are the tool kits, frameworks, or mechanics that can restore balance. Reinventing the rules when working in areas outside of and sometimes against convention requires careful and intentional consideration. "The idea of co-creation, collaboration, equal partners, that's not something you learn in school," said Julia Kumari Drapkin of ISeeChange, "not [at] the journalism school I went to."

lessons from field

01   deep listening and dialogue

02   develop common principles

03   consents, community benefit agreements

04   process in balance with outcomes

05   complex narrative structures

06   healing, safety and sustainablility

07   appropriate forms of leadership, language and technology

08   media and digital literacy

09   alternative models of funding, evaluation and impact

10   iteration

Lessons from the Field. *Courtesy Co-Creation Studio.*

Importantly, the following suggestions are not a to-do list, nor are they prescriptive formulae. Rather, these are areas of consideration and practices that have worked in specific contexts, cited to help begin discussions and prompt other adaptations to local and unique situations. "The toolkit phenomenon is not new," Bangalore community organizer and facilitator Vishwanath (Zen Rainman) said in a conversation with Babitha George. "But one has to also consider societal aspects here. . . . There are no generic solutions. There has to be localization of knowledge, inputs and experience."[8]

Fred Turner, who famously connected *Whole Earth Catalog* counterculture with cyberculture, worries about what a limited prescriptive the tool kit format can imply:

To the extent that we imagine that politics take place in the intimate realm of personal power, we're going to get lost. We're going to keep building interfaces that allow for expression, that allow for the extension of intimate personal power, and we're going to precisely not do the work, the boring, tedious, structural work of building and sustaining institutions that allow for the negotiation of resource exchange across groups that may not like each other's expressions at all. So we have inherited from the *Whole Earth Catalog* a language of individuals, tools, and communities, which we've translated, I think, in tech speak, into individuals, communities, and networks. I would like to see a language of institutions, resources, and negotiation take its place.[9]

LESSON 1: GATHERING, LISTENING, DIALOGUE

How do co-creative projects begin? A conventional feature film, art exhibit, or public-service campaign will have predefined outcomes, and the makers will seek out relationships, crew, subjects, technologies, resources, and methods to serve the goal. Instead, co-creative projects begin with relationships. These projects start by creating a space wherein people may gather and ask questions. There may be a complex problem or set of problems that people want to address. Eventually there will be outcomes, often many, but these emerge out of the process rather than the other way around. Co-creative relationships begin by listening.

"Pathways to participation" is how the filmmaker Michael Premo described to us this early stage of process at Storyline, his organization with Rachel Falcone. "We convene open meetings by reaching out to groups, organizations, friends, colleagues, people we admire. Sometimes those conversations are one-on-one, so we'll just talk to people directly and be like, 'So, what's happening? What are you up to? What campaigns and things [will] you look at in the next twelve to eighteen months?'"

Peruvian women share their stories of forced sterilization under former-president Alberto Fujimori in the 1990s. From the Quipu Project, *an interactive documentary co-created with Indigenous women and local community organizations. Used with permission from María Court and Rosemarie Lerner.*

Most co-creators identify deep listening and dialogue as not only the key first step but also a continuing essential element in the process of developing joint and equitable project designs. As Gloria Steinem wrote about listening, "One of the simplest paths to deep change is for the less powerful to speak as much as they listen and for the more powerful to listen as much as they speak."[10]

Many veterans of co-creation equate the process with a loss of ego accompanied by a good dose of humility. Judy Kibinge, who co-created Docubox, an organization dedicated to serving the documentary community in East Africa, said, "I think it begins with recognizing that you really don't know anything. . . . You don't know anything at all. And so, you create a program and you try to be very receptive to the needs of the people that you created the program for."

Vincent Carelli of Video nas Aldeias in Brazil described this as "countermethodology. . . . It's about going to people's houses, building bridges, and lots of listening. It's not a technical thing. It's about discussions that are not imposed. So, the most important thing is the dialogue."

Vincent Carelli of Vídeo nas Aldeias in Brazil described the organization's work as "countermethodology." Used with permission from Vincent Carelli, Vídeo nas Aldeias.

In his research for a book on designing dialogue, Fred Dust, formerly of IDEO, spent time at Quaker meetinghouses to learn about consensus. "Quakers believe that decisions will come out of the silence of listening, and it's a collective process," he said. "It's not exactly storytelling, but it is an act of co-creation."

Of course, such attempts at eliciting community input can still be prone to the hazards of exploitation and hierarchy, in part through what place maker Jay Pitter acknowledged as the "narrative risk" in the process. "One of the ways that social imbalances manifest is that historically marginalized people are often required to share delicate parts of their stories, often rendering them more vulnerable. Individuals with more social power tend to perform their stories with omissions that protect their social status and power. We rarely explore the inner worlds of individuals from socially powerful groups; they are usually the ones on the safe side of the storytelling process. I always say as a starting point that you actually don't have a right to ask a question . . . that you wouldn't answer yourself."

The first stages of co-creation may be open but not completely without structure or facilitation. The process is a continuum, said civic media scholar Ethan Zuckerman. "There's the orthodox method, where you really start with that blank sheet of paper. But then, particularly around technologies, which can sometimes bring capabilities to the table that are nonobvious, I think where we're starting to end up is iteration, where we are putting out that sacrificial technology, building a community, contemplating what might happen around it and then iterating and maybe changing who's at the table, maybe changing what the prompts are."

Michael Premo identified such prompts as "handrails" to help give a framework to the discussion rather than leaving it completely open-ended.

One group has developed a tool kit specifically to help guide how human rights defenders and affected communities might work together. Engage Media and members of the Video for Change network published the online "Video for Change Impact Toolkit" in 2019.[11]

Other co-creative teams have created decks of playing cards with prompts that offer nonlinear stimulation for dialogue and conversations. The Iyapo Repository is a (currently itinerant) "future museum" and resource library that, according to its website, "houses a collection of digital and physical artifacts created to affirm and project the future of people of African descent." For its community outreach and media-literacy sessions, it created such a deck of these cards because "[they] realized there needs to be some scaffolding," according to one of the group's cofounders, artist Salome Asega. "It is actually impossible to just get people to think about the future without directions,"

The Iyapo Repository created a deck of playing cards to use during participatory media workshops. Used with permission from Salome Asega.

she said, "so the cards provide prompts and scaffolding. The conversation can go anywhere, but we need the things to ground you, to give you a starting place."

Similarly, Jessica Clark of Dot Connector Studios created a deck of cards she uses to inspire ideas for media-impact strategies, specifically for funders and journalists. "There's no one linear path—it's a mix-and-match moment," she explained.

Listening does not end after the first meeting of a co-creative team and requires a serious commitment to iteration the part of the team and agreement that things will change along the way. Too often, members of projects stop the listening portion of the project after the first round and later, reducing the approach to a onetime consultation or even data mining by those in control. But that is where co-creation ends and the problems set in. In authentic co-creation, these initial steps are only the first term-setting conversations and by no means the last.

LESSON 2: MANIFESTOS AND STATEMENTS OF PRINCIPLES

Vertov's Kino-Eye collective made a manifesto, as did the early hackers and the "Mundane Afrofuturists."[12] Manifestos are declarations of the ethics, principles, and values of groups that push new ideas. Many of the co-creators, collectives, and media movements in this study have assembled similar lists of principles. Allied Media Projects, Data for Democracy, and Platform Cooperativism are only a few of the dozens of organizations that have published sets of common values that define and guide their work. The Design Justice Network goes further to document and make its processes transparent through publishing its elaborate ideation sessions as well as a wiki version.

In 2005, Katerina Cizek and her co-creators drew up a manifesto in the early years of the NFB of Canada hospital-based *Filmmaker-in-Residence* project. This was a first step largely used as a tool for communicating with potential partners and co-creators, who often assumed that the team simply wanted to make films about them. The manifesto spelled out a different intention: to work with health care workers as well as communities in a joint process of discovery rather than documenting what they already knew. Cizek and the team adapted a similar manifesto for the *HIGHRISE*

Film still from Soviet filmmaker Dziga Vertov's Man with a Movie Camera *(1929). Vertov was the founder of the Kino-Eye filmmakers group, which penned one of the earliest cinema manifestos. Image printed under fair use.*

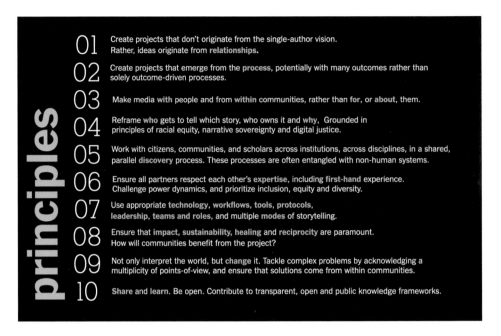

The Co-Creation Guiding Principles. Courtesy Co-Creation Studio.

project (2009), which outlined the common principles shared by their many co-creators: architects, residents, government, technologists, and artists. The manifesto that we have developed for the Co-Creation Studio at the MIT Open Documentary Lab is based on those two earlier documents, as adapted through discussions with focus groups and others.

For *SGaaway K'uuna* (*Edge of the Knife*, 2018), the Haida film project (see chapter 3), the first thing the makers did was draw up a list of goals and values. They ranked these and found that the highest priority item was "authenticity," giving the group an imperative to represent Haida culture accurately and always in close consultation with elders and knowledge holders. According to Jonathan Frantz, the producer and cinematographer, "Then further down the list were more conventional things, like make a film that has high production value or . . . might have a chance to make it in a film festival. . . . Anytime we had any roadblock or stumbling block with decision-making, we would go back to our set of values and use that to help us make a decision on which way to go."

Naming common values helps identify what matters most to group members and articulates their reasons for coming together.

On the set of SGaaway K'uuna (Edge of the Knife). *Photo credit* شون حرف *Farah Nosh. Used with per-mission from Isuma Distribution International.*

LESSON 3: CONTRACTS AND TERMS OF ENGAGEMENT

This is the paperwork. While some co-creative teams never sign anything, many use negotiated documents up front as a way to articulate and agree on relationships, ownership claims, and ultimate benefits. The standard tools in film and media include the individual consent or personal release forms. Filmmakers ask subjects to sign forms stating that the participant agrees to be in their films and are releasing their image, words, and sounds, often in perpetuity, for all media everywhere. In academic and medical research involving human subjects, this is known as individual informed consent, purporting to ensure that the subjects have a basic understanding of the research; consent to releasing their information and data, often with a guarantee of anonymity; and have the right to opt out at any time. The inconsistent, troubled history of the use of these tools dates back to 1900, which is considered to be the first time informed consent forms were used in medical research by the US Army in studying yellow fever.[13]

Some interdisciplinary co-creative projects have attempted to reconcile the opposing guarantees of media releases and research consents. It is difficult to guarantee anonymity and a right to withdraw from a project when media is involved. In the digital era especially, images, text, and sound, once captured, can be disseminated anywhere with or without consent (sometimes even that of the media makers). Louis Massiah of the Scribe Video Center challenges the standard that consent forms be signed before an interview, as most academic and broadcast protocols dictate. "It feels a little bit presumptuous, if not downright colonial and colonizing. . . . I look at an interview as a gift the witness/interviewee is sharing with the media maker, as opposed to selling something to be commodified. When you make a community media work, the people that are on camera or participating as subjects are doing it because they believe they are trying to support the goals of the project, and they also need to understand that the edited work is going to be in their own interest."

Used well, such documents can be an opportunity to conduct a deep dialogue between participants and coresearchers and offer a range of options that details the risks, responsibilities, and benefits for all involved. In the NFB *Filmmaker-in-Residence* program, for example, when working with a group called Young Parents of No Fixed Address, Cizek and her team created iterative and staged levels of informed consent with corresponding levels of anonymity and participation. The health care–based research coordinator developed the tools and then spent hours discussing options with participants, often young people at risk of being harmed by going public with their stories. The team returned to these principles and options frequently as the projects developed.

In *Dimensions in Testimony*, Stephen Smith of the Shoah Foundation worked with Holocaust survivors in a new technology environment. He noted that it could be daunting, ethically, to place "an 85-year-old individual into a 15-foot dome with 6,000 LED lights and 116 cameras and ask them a thousand questions about their life over five days. Just seeing that, that would be an imposition and possibly dangerous for that individual to go through that experience."

Smith, who holds a PhD in testimony methodology, developed a complex protocol with his team and with survivors who sat on an advisory board to shape the process. They asked that family members rather than psychologists or social workers accompany all interviewees to act as advocates.

A photoblogging project with members of the group Young Parents of No Fixed Address featured in the NFB Filmmaker-in-Residence program (2006). Used with permission from the National Film Board of Canada.

With Dimensions in Testimony, viewers have conversations with Holocaust survivors and other witnesses to genocide through interactive recordings of interviews. This approach allows viewers to create their own interview experience by directly asking the questions they have for survivors and receiving testimonies from the survivors, mediated through interactive techniques. Used with permission from the USC Shoah Foundation.

Many co-creators have also identified the need for instruments that help address inequities at a more systemic level and guarantee rights and responsibilities to entire communities. In 2009 Screen Australia, a federal media agency, published the influential book *Pathways & Protocols: A Filmmaker's Guide to Working with Indigenous People, Culture and Concepts*. Therein, the Indigenous writer and lawyer Terri Janke offers "advice about the ethical and legal issues involved in transferring Indigenous cultural material to the screen."[14] A similar document in Canada was published in 2019 called *On-Screen Protocols & Pathways: A Media Production Guide to Working with First Nations, Métis and Inuit Communities, Cultures, Concepts and Stories*, which helped pave the way for the subsequently formed Indigenous Screen Office.[15]

Ellen Schneider of Active Voice has developed customized tools called "Pre-Nups for Filmmakers." She has worked with funders, filmmakers, and nonprofits to navigate the increasingly complex nuances of co-creative partnerships that go beyond the mold of conventional filmmaking. "When I first started working on this, some people were concerned that even raising those what-ifs could be daunting and a deterrent," Schneider told us. "But the more that we dug into it and the more people gave us recurring feedback, people said, 'Let's just get it on the table, because if there's concerns about, you know, shifts or changes of plans that we have no control over, let's deal with it now.'"

Schneider has broken down the process into four sequential stages: "mission, method, money, and mobility." The big stumbling block can be money, followed by how the project will move in the world.

"Who's controlling what resources and how are they allocating them?" asked Sheila Leddy of the Fledgling Fund. "Many people think of accounting as just financial accounting, like your income statement and balance sheet. But there's a broader understanding of management accounting that is very much about thinking about cost drivers and who has control over those and how you're reporting on those so that you can manage them, not just report out. So, what are the systems that are going to be in place, and who's included in actually figuring out what that budget is?"

How will decisions get made and at what stages? Sasha Costanza-Chock, a civic media scholar and designer, said that this is key to a co-creative agreement. "Is it consensus? Who decides?" In the software world, Brett Gaylor, a documentary maker formerly with the Mozilla Foundation, noted that teams working on projects will specify "who is responsible, accountable, consulted, and informed," and the model is now used in digital agencies and increasingly in interdisciplinary emerging-technologies storytelling.

Meanwhile, cultural groups are exploring cultural community benefits agreements (CBAs) that spell out the terms of engagement, especially for outside organizations coming in to work with local groups. Indeed, the Detroit Narrative Agency (DNA) asked the MIT Co-Creation Studio to negotiate a CBA around the terms of our mentorships and coresearch with them. This model picks up from the CBAs that have become common in urban contexts in recent decades to negotiate terms for local communities when developers arrive to transform neighborhoods. These are legally binding instruments signed by developers, municipal governments, and community groups. The benefits at issue might include local jobs, living wage requirements, affordable housing, and neighborhood improvements, according to the Parkdale People's Economic Project.[16]

These types of agreements date back even further in Canada's north, specifically with the mining industry in the 1980s. There, they are known as impact and community benefits agreements. However, critics warn that they can have an exploitative effect by, for instance, compromising communities' land claims. They are not public policy documents but instead are confidential agreements subject to contract law and often demand that communities relinquish the right to protest or even publicly criticize the companies in question.

CBAs have also been used as part of the amalgamation of the broadcast system in Canada. Ana Serrano, former director of the Canadian Film Centre, notes that her organization profoundly benefited from this process, which mandated that a portion of broadcasters' taxes went to community groups such as hers. But the effects can be fragile. She told us that throughout her "career at the Film Centre, it was one of the big ways that cultural institutions got money. But in the past five years, the community benefits were taken out of these deals."

Consent forms, community-benefit agreements, and memoranda of understanding are not panaceas. Veterans of co-creation are acutely aware of their pitfalls and problems but do use them to guarantee certain basic rights, alongside bigger policy and legislative concerns affecting the role of government; this includes how the commons will be defined and how communities will govern themselves and their shared resources into the future.

IN CONVERSATION: THE BENEFITS OF COMMUNITY BENEFIT AGREEMENTS

by DETROIT NARRATIVE AGENCY

Edited excerpts from a group interview in which members of the Detroit Narrative Agency discuss the importance of putting terms in writing before co-creative work begins.

ill Weaver of the Detroit Narrative Agency (2018). Photo by Kashira Dowridge. Used with permission of Co-Creation Studio.

ILL WEAVER: For DNA, when we first got approached by MIT to collaborate on this research process, we wanted to make sure that the process felt really equitable, not just to us but, more importantly, to the cohort members and any other community members that would interact with the process. For us, it felt really important that there was a set of clearly stated not just principles and guidelines but actual expectations around how the process will play out, to have some tangible ways that we can strive towards and measure how equitable the process would be. So then we invited the MIT team from Co-Creation Studio to join us in negotiating a working document that would be a cultural community benefits agreement in the form of a memorandum of understanding between DNA and them.

PAIGE WATKINS: . . . I think that you run the risk of whatever is produced—if it's an event, if it's a conference, if it's a gathering of people, some kind of institutional partnership—I think you run the risk of spending all this time planning, and then the thing happens and it's not what you agreed to. Or they last-minute changed all this stuff around about it, and now all of a sudden it's out of value alignment. It could be exploitative to the community members especially in Detroit, a place where more and more people are wanting to have conferences here, more and more people come here to do things, thinking that they can just kind of come in and do their thing and then leave without being accountable to anybody.

So I think you just run the risk of folks doing whatever the hell they want and not actually thinking about what the community needs or thinking about actually supporting folks from the community. You [could have] an event that was like a couple-hundred-thousand-dollar budget, you know what I mean, and all of these vendors are White folks from the suburbs or something like that. You just run the risk of things happening that are out of alignment or out of whack and then having to backtrack and fix stuff rather than doing things in the beginning before anything's decided.

ILL WEAVER: And I'll just add to that, that you know I think even in the best-case scenario where people have similar values at institutions and they come in with really good intentions and all that, then like if they're not clearly articulated and you're not building that in as a structured formal part of the process, then you run the risk of your intentions falling through the cracks because the logistics end up trumping and the economics end up trumping your politics or your values, you know what I mean? What's the thing Grace Lee Boggs would always say? "Politics in command." It was a phrase that she used to say about how we end up running the risk of our politics falling to the wayside to make certain economic decisions or logistical compromises, because there's systemic exclusion and divestment from our communities. So then, how do we proactively do that?

And that's on a money-spending level, which is easier to track. But then I think we're also interested in the percentage of projects that get featured in an event or the background of the trainer that we collaborated with . . . I think to prioritize Black and people of color filmmakers so we're not replicating this thing of outside-of-Detroit White people as the experts of this storytelling process. So, these are all part of the equity piece. I think in other projects we've also looked at safeguarding against co-optation, and I think that's something that we really thoughtfully did together with MIT in our CBA with them: How do we make sure that DNA is listed as coauthors, and there's input at every step of the process around content? And that's not to make any assumptions or accusations that things will go wrong, but it's just about how do you build a process that gives you agency and space throughout the process to chime in if there's a red flag or a concern?

MORGAN WILLIS: I think the last thing, what you all are also making really clear, is that community benefits agreements have the opportunity to invest in local work and local people who are doing work that allows their platform to grow. So, it's also a way of tangibly investing in our local economy. Then folks who are here have an experience that is more reflective of the actual local community. They also have reference points of "Oh my God, did you do this thing? Did you taste this person's catering?" or whatever. We get really excited about whoever the person is or the place is because it's gonna be something that they've not had before. Typically they've only relied on online searches to tell them what their nearest, Whitest, or most resourced people who do this work are telling them to do.

LESSON 4: BALANCING PROCESS WITH OUTCOMES

Many co-creative projects directly challenge the industrialized, product-driven, commodified notion of media making. However, there comes a time to discuss the media itself: Just how important is their quality?

The NFB's Challenge for Change program fifty years ago prided itself on being profoundly focused on process over product. This was radical for a film agency that had been, at that point for thirty years, at the vanguard of championing documentary and animation as legitimate art forms and industries. Although the Challenge for Change program ran for a decade, much of the thousands of hours of resulting film material was never edited or synthesized but instead was used only in fragmentary ways for community engagement. Of course, some films were completed, notably the controversial *The Things I Cannot Change* (1967), which depicted its subjects, an impoverished family, in an unflattering light. As a result, George Stoney, head of Challenge for Change at the time, wrote a harsh treatise demanding a new code of ethics for the program.[17]

Colin Low, director of the Challenge for Change Fogo Island project, later recalled his 1971 debates about process versus product with NFB founder John Grierson a mere three months before Grierson's death. Over the course of three guest lectures in front of students in Grierson's university classes, the two battled over the value of the Fogo project and issues of journalism, propaganda, and editorial style. As Low recalled, "'What,' Dr. Grierson wanted to know, 'was the value of the film off Fogo Island? Was it good for television? Mass media? What did it say to Canada? . . . Why would the Canadian taxpayer allow such an indulgence?'"[18]

Low was devastated by Grierson's exasperation with Challenge for Change. Grierson was craving drama, while Low had thought he was fulfilling Grierson's own call to make "peace more exciting than war." Thirty years after this debate, the NFB sought to revive Challenge for Change but this time aligning process with product.

Workers at the Fogo Island Co-Operative, in a scene from one of twenty-seven films by Colin Low about life on Fogo Island, Newfoundland, which was part of the NFB Challenge for Change program. Used with permission from the National Film Board of Canada.

When Katerina Cizek and a cohort of filmmakers were invited to reinvent the program in the digital era, the producers, Peter Starr and then Gerry Flahive, attempted to embrace the ethos of Grierson and Low. The producers asked that Cizek put process first, add layers that would create media that would extend beyond the communities it first served, and speak to broader audiences. Initially the NFB wanted a feature film about the process, which Cizek worried was too laborious and intrusive, but then the team went on to consider the web and to co-create the world's first online feature-length documentary for global audiences, *Filmmaker-in-Residence* (2004). Along the way, they decided to produce multiple products that would serve different needs. Short, rapid community-facing outputs such as teaching tools, short videos, newsletters, and flyers were sometimes more important to community partners than the high-quality media product intended for wider audiences.

For many groups in this century, both the quality of the process and the quality of the production are crucial. For decades, community-based media work has been relegated to low-quality aesthetics and unrefined narrative structures, but today, when media tools are more accessible and high-quality visuals and narratives are ubiquitous, co-creators also see the need to create attractive and engrossing products. As paige watkins of the DNA commented during a group discussion, "If you want the product to actually have impact past the choir, past the people who already understand what we're talking about and are agreeing with our values[,] . . . it has to be competitive up against the harmful things that are getting maybe more money or more resources."

"Projects emerge from the process," is how Heather Croall, director of the Adelaide Fringe Festival, succinctly explained co-creation in the twenty-first century. For most co-creative teams, it's not one or the other. Product and process are complementary.

LESSON 5: COMPLEX NARRATIVES AND BEYOND STORY

When projects are born from co-creative processes, they often flower into diverse, alternative forms of narrative structures. Co-creators can shed linear, conventional formats (such as three-act structures) and embrace open-ended, ongoing, multivocal, and circular, spiral forms. They can take the documentary "beyond story," as suggested in a recent collection of essays critiquing the industry's penchant for three-act, hero-centered, and prescriptive story forms.[19] These co-creative innovations can challenge conceptions of quality to widen beyond traditional aesthetic norms. "There's potential, for the form itself, that co-creation can result in an expansion of the language of nonfiction," said Tabitha Jackson, director of the Sundance Film Festival.

"Rather than deductive argument or grand narratives, these projects employ mosaic structures of multiple perspectives," wrote Patricia Zimmermann and Helen de Michiel in *Open Space New Media Documentary* "Open space new media documentaries explore the terrain where technologies meet places and people in new and unpredictable ways, carving out spaces for dialogue, history, and action."[20]

Complex, narrative world-building and story-universe labs proliferate in science-fiction and speculative-media spaces. Alex McDowell's World Building Media Lab at the University of Southern California and Lance Weiler's Digital Storytelling Lab at the Columbia University School of the Arts each co-design spaces with and for participants to imagine future worlds. In McDowell's world-building philosophy, the design of a world precedes the telling of a story.

While science fiction and documentary nonfiction might at first seem worlds away from each other they can coexist, especially in a co-creative cosmos. In 2020, the Guild of Future Architects began running Futurist Writers' Rooms that encourage

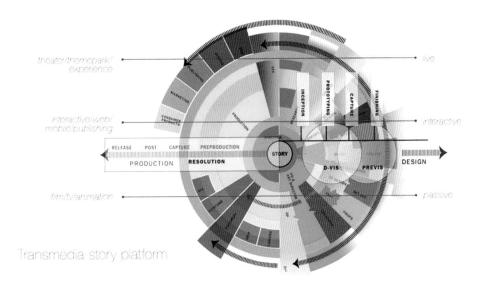

Graphic from worldbuilding.usc.edu. The World Building Institute illustrates the complex iterative process of creating story worlds. Used with permission from Alex McDowell.

participants to "imagine the past and remember the future" by collaboratively creating alternate histories and speculative futures for human systems, according to Kamal Sinclair, Guild of Future Architects executive director. The guild's co-creative methodology aims to help teams forge deeper innovation and design in media, policy, and planning across many sectors.

In the 2019 interactive documentary theater piece *Casting the Vote*, by documentarian June Cross and theater director Charlotte Brathwaite, the audience and actors co-created the performance every night, embodying the concept that democracy unfolds in real time. As participants shared a meal alongside the performance, they were informed of an omission in the story being told. "Chef Mea Johnson from the Olio Culinary Collective, a Native American from the Wampanoag nation, designed the meal to honor Native American and African-American traditions," wrote the piece's director, journalist June Cross. "At dress rehearsal, she pointed out that the film we'd edited omitted mention of how Native peoples in North Dakota had been prevented from voting because they have post-office boxes instead of street addresses. We asked her to add those comments as she explained the meal."[21]

By incorporating new voices into the performance at any stage, *Casting the Vote* embodied democratic principles of inclusion in its storytelling and complicated its many layers of narrative.

Open narrative forms can be deeply connected to reimagining civic engagement. In *For Freedoms* (2018–present), Hank Willis Thomas makes a direct connection between art and civic life by creating town halls and large public art projects in each of the fifty states. He told us that "everyone has something to say and something valuable and something important or interesting. It's just about what best suits them. The more people, I think, are invested in making space, new stories, and new ways of telling stories, the better off we all are."

Henry Jenkins's Civic Imagination Project scales this kind of approach as an ongoing action and research agenda at the University of Southern California, based on eight years of research. The project taps into the speculative capacities of citizens at the local level, bringing together and bridging the perceived gaps between different stakeholders, especially youths. The team conducts research and ethnographies and runs workshops across the country to seed trust and new imagined civic futures.

Most interactive and immersive documentaries challenge the conventions of linear, three-act storytelling. They are often multivocal, sometimes characterless, and open in the way that audiences can engage with them and chart their own way through the experience. Many times, as Amelia Winger-Bearskin says below, documentaries can take complex shapes and contours within Indigenous ways of knowing.

This billboard by artist Amelia Winger-Bearskin was installed at the Anchorage Museum in Anchorage, Alaska, in 2020, as part of the For Freedoms *project. Photo by Jim Kohl. Courtesy of Amelia Winger-Bearskin.*

IN CONVERSATION: DECENTRALIZED STORYTELLING

by AMELIA WINGER-BEARSKIN

Excerpt from a conversation with Amelia Winger-Bearskin, artist and curator, with text added with permission from her writing.[22]

AMELIA WINGER-BEARSKIN: We [Indigenous peoples] have our own traditions, and I just want to make sure we're able to speak about them, learn about them through the context that we give them. And I'm certainly someone that believes in sharing our culture and having other people participate with it, but also, I want to make room for us to have our chance to say what we want, to say about our culture and then also contribute to other research.

One of the main concepts I've been working with is decentralized storytelling. I'm charting the history of the type of storytelling that I'm seeing that's most popular among younger millennials, that's created through different online platforms. . . . They take the form of VR Chat and Rec-Room and Alt-Space and indie games and *ROBLOX* and *Minecraft*, and I can go on and on and on. It's [for] ages from ten to maybe twenty-five, and they are telling long-form, complex narratives from hundreds of people's perspectives that are all participating in a story. And that is a more similar corollary to the way in which the Iroquois would tell stories.

The Iroquois with Six Nations spanned a massive amount of area in the Northeast. They had trade routes all the way down to Chile and beyond. So, in order to share the technological advancements that they made as well as the spiritual lessons that they were learning from different communities, they would do it in the form of story. The stories were very long-form. They would use a multiplicity of different types of tools—pottery, baskets, rock formations, natural land monuments—as both mnemonic devices, to remember details about them, as well as to share those with other groups of people.

These were very complex and long stories—[it] would take a week to tell the section of the story that you're trying to convey. It was entertaining, it was community building, it shared values, but it also shared technological advances from one community to the other by connecting to a story that people were familiar with. It gave a wider chance of being remembered and being relayed.

For instance, we still make cornhusk dolls. The cornhusk doll story is a way of teaching a specific type of planting process and harvesting process. And the Iroquois farmers taught Benjamin Franklin how to farm and taught him about decentralized systems of democracy and a model for how agriculture was created within the

Corn husk doll. Photo by John Morgan. Distributed under Creative Commons Attribution 2.0 Generic License.

United States. And so . . . the reason you use a doll to tell a story is that a child learns and remembers, and you have a longer chance of having that community remember the story. If someone learned that story when they were five and they remember the story when they were six and kept hearing the story as they grew up and they begin to plant, they remember this technological [lesson].

So, that's just an example of how stories were distributed across hundreds of people. They also had different formats of story: it was a doll, it was something that

Wampum.codes is Amelia Winger-Bearskin's project to develop a model of ethical software depen-
dencies, an attempt to inscribe community values and developer accountability into code. Image
courtesy Amelia Winger-Bearskin.

someone would say, it was a song. . . . And it was also a time of year. And that, to me, seems more similar to the way in which the younger generation is telling their stories, through these distributed experiences.

Decentralized storytelling requires navigating your human experience within a story space. The architects of the story space were your ancestors; it is a place in which your great-great-great-great-great-grandchildren will be born. It is a mind space, a physical space, and, for some, a game. It is a process of making and unmaking that cannot be sustained except in collaboration with the past and future, securing a connection with those living seven generations from now, listening to their experiences, and making stories that will be able to reach them where they are. A decentralized story gets its life from the people who participate in it. This is one reason that decentralization flourishes so well in the worlds of memes and fanfic—it's easier to put on puppet shows with familiar characters than it is to create a whole new fictional universe from scratch. By using ready-made features of an existing cultural landscape and doing it on a platform that everyone can access, with tools that everyone has, decentralized stories make it possible for more people to participate in the telling of the story.

LESSON 6: SAFETY, HEALING, AND SUSTAINABILITY

Safety first. What does it mean to prioritize the well-being of the people involved in co-creation projects? It means moving projects and participants toward healing and also thinking profoundly about the sustainability of the project by protecting its participants and building and leaving renewable resources within the community. Bringing people into co-creative processes can make them vulnerable whether in person or online, especially in an era of toxic social platforming. As Brett Gaylor, formerly of the Mozilla Foundation put it, "Bringing people in, they're not necessarily in our tool kit. Having concrete methodologies to do that, I think, can be really transformational, and it also protects the safety of everyone involved, and I think that's a real critical thing that's needed in the field."

Too often, the most vulnerable actors are the ones working independently at the local level. As of April 2022, 1,440 journalists have been killed worldwide since 1992, according to the Committee to Protect Journalists,[23] and many other activists are also under threat. Along with the work that WITNESS does in helping people produce and safeguard video evidence of human rights abuses, program director Sam Gregory is exploring the idea of "co-presence," which involves livestreaming well-informed witnesses from remote locations into crisis areas, such as protests and stand-offs. As Gregory told us, "Witnessing implies the obligation to watch and to act. What is it we can do that involves bringing people into livestream that makes them active agents in the narrative as it goes on, driven by the needs of the frontline activists? Not to tell people on the front lines what to do, but to do things that could change events by the support . . . [,] for example, by translating or adding context or sharing it rapidly. Or can they provide specific skills, for example, a legal observer?"

Many communities involved in co-creation have experienced oppression and trauma. At the Dalit-led South Asian American arts, organizing, and technology collective Equality Labs, founder Thenmozhi Soundararajan and the team have integrated digital security as a core element in their artistic and activist work to "organize safely and protect ourselves and our communications," with training workshops, resources, consultations, and a rapid-response team.

Unfortunately, the very process of co-creation can expose, reopen, and even create wounds. As Soundararajan noted in her 2021 talk at our lab, "While we are digging through the bones, the forensics of the violence incurred upon us, our oppressors are stealing the stars."[24] How, then, do teams not only mitigate trauma but also actively contribute to reconciliation and collective health?

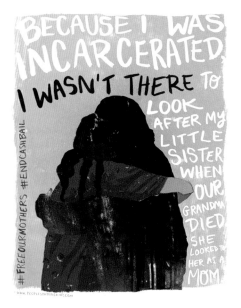

Illustrations from the People's Paper Co-op, which is focused on healing, reentry, and sustainability. Used with permission from Mark Strandquist.

Loretta Todd, a Cree/Métis filmmaker, highlights the crucial role of ceremony and protocols of media making within Indigenous communities. In the *On Screen: Protocols and Pathways* report, she's quoted to that effect:

Again, if we're trying to decolonize the documentary, then we can bring in our own forms of how we talk and how we do business and how we deal with one another, and one of the ways we deal with one another is that we honour one another, and that often comes through song. Whenever I've been to meetings or a ceremony or a cultural gathering, there's usually a welcoming song or honouring song for the people who are participating. That honouring song honours the people who are sharing their stories, and honours the people who have arrived to hear the stories. That honour song hopefully creates a respectful space to start the film off so that the people who are there will listen and open their minds to the stories they are about to hear. It's also my way of honouring the people who have shared their stories with me to make the film. I'm trying to bring those ways of how we do business into the practice of filmmaking.[25]

In *Emergent Strategy* (2017), adrienne maree brown advocates for approaches that are informed by transformative justice that "looks for alternative ways to address/ interrupt harm, which do not rely on the state[, and] relies on organic, creative strategies that are community created and sustained."[26] Healing also involves sustainable

A still from Fireflies: A Brownsville Story. *Used with permission from the Brownsville Community Justice Center and Peoples Culture.*

development of a project and its communities. Further, many co-creative projects are interested in creating renewable economic resources for communities. The Brooklyn-based project *Fireflies: A Brownsville Story* is not only a documentary VR game but is also a peace-building process that allows residents to cross territories virtually that are normally off limits to them physically. The Brownsville Community Justice Center and Peoples Culture installed the game in the lobbies of more than fourteen public housing buildings to nurture conversations about how to heal the fractures in the community. The project maintains the Brownsville Tech Lab in order to train youths in game design, coding, and other crafts of media making. As producer Sarah Bassett told us, they "want this game to live in the world, but at the end of the day it's made by and for the community with the purpose of [being] shown and played in the community."

LESSON 7: APPROPRIATE LEADERSHIP, LANGUAGE, AND TECHNOLOGY

APPROPRIATE LEADERSHIP

In the twenty-first century, conventional notions of employment and jobs are transforming in a massive business-driven paradigm switch. In rebuilding a postpandemic society, artists need to be understood as leaders of collective healing, argues Marc

Bamuthi Joseph in the Ford Foundation report *Creative Futures*. "I believe we must deploy artists as leaders as we collectively heal from the trauma of the pandemic. Without inspired, cohesive political leadership, it'll fall on cultural leaders to design and model our post-COVID-19 healing apparatus. Artists must help us re-learn to feel alive in public."[27]

Sheila Leddy of the Fledgling Fund has worked with filmmakers to help them adapt to changing systems. "When Fledgling first started doing this work," she said, "the business models were changing, and expectations were changing. There was pushback about impact: 'What do you mean I have to care about impact? What do you mean I have to measure this? I tell stories, I'm a filmmaker.' . . . But I think storytellers need to be part of the conversation and to participate in the development of new models, or you're going to get run over by the business people."

What do new models of leadership look like with a mindful, intentional approach to creating new spaces for co-creative models? In flattening and shifting power dynamics, co-creators learn to respect expertise in nonhierarchical systems. In nonfiction media, this means that the people formerly known as directors and producers take on new roles. We asked in our interviews and group sessions what these new roles could be called.

"There's a certain conception of what a leader is: the lone figure. It's like Gandhi, right?" said Lina Srivastava of Creative Impact and Engagement Lab:

Graphic by Katerina Cizek. Used with permission from Co-Creation Studio.

[Gandhi] was indeed one kind of a leader—he did lead a national resistance movement—but there was an entire mechanism that was working alongside him. He did nothing alone. Take the Salt March, for example. The movement was a very strategic movement. There were multiple parts, and it was leaderful. I hate that word, but it was leaderful and collaborative. It's the kind of leadership that catalyzes transformational change in communities and in societies. So, the important questions for me are [these]: How do you apply leadership process to any particular social, cultural, political context? What are people and communities doing together, collectively? The answers are all based in stories.

Srivastava founded the Transformational Change Leadership project to document and communicate those stories.

"I've been a part of lots of different communities," said Eleanor Whitley of the pioneering immersive story studio Marshmallow Laser Feast. Those included "intentional communities where these kinds of leaders suddenly go 'I don't want to do this anymore. It's too heavy and it's too difficult and it's all of you have to decide.' And you're like 'No, actually you do need to—however you design what that community looks like—to create a path and to guide people down.' Leaderless teams fail."

Leaders might also consider themselves more instigators than primary producers. In projects such as It Gets Better and the Candles 4 Rwanda campaign, it was a matter of creating a template that a wide range of other media makers, professional and amateur alike, could fill in with their own video messages and provide a platform to disseminate these varied voices.[28] "Co-creation . . . is a means to giving people what they need," said filmmaker and media strategist Jennifer MacArthur. "And sometimes what people need is less power, to give them more of a sense of their humanity. And it's hard to know that, and it's hard to accept that in a culture that is obsessed with 'bigger is better.'"

For Wendy Levy, director of the Alliance for Media Arts+Culture, intergenerational collaborations in particular can suffer from a lack of relinquished power. "'Youth media' and youth makers are often marginalized, excluded or denied access to decision-making in supposedly co-creative environments," she told us. Levy suggests that creating shared language, understanding creative leadership models, and developing custom methodologies for specific projects can challenge structural inequities and help decentralized leadership.

APPROPRIATE LANGUAGE

Language can unite and can also exclude. In co-creative processes, when people from all walks of life come together to listen, engage in dialogue, invent, and produce, the different vernaculars of their respective specialties and communities can quickly get in the way and reproduce hierarchies in groups.

As Heather Croall, former head of Crossover Labs and director of the Adelaide Fringe Festival, told us, "You learn a shared language, [but] when you've got people from different sectors, . . . everyone talks in a very different language, so coders talk in a very different language to game designers, to filmmakers, to scientists, to artists. You realize that most people walk in the room with their arms crossed and they're not really all that open to talking with each other. And then by the end they're really into it."

Some co-creative teams develop glossaries of terms and catalogs for their time together to develop both mutual understanding and an efficient shorthand. But a simple watchword might suffice: "Talk normal," said Fred Dust, author and former partner at IDEO, "which seems easy, but it's not. We as designers, we as filmmakers, we as other kinds of practitioners, we learn special language, and that actually can be highly exclusive language. So, learning to just talk normal is really foundational to actually being a co-creative body and actor."

APPROPRIATE TECHNOLOGIES AND TOOLS

New technologies are exciting, and uncharted territories allow for co-creation to flourish. "You're not in a templated medium when you're building something new," said Liz Rosenthal, a member of the London-based film and media consultancy Power to the Pixel and cocurator of the Venice Film Festival's VR Island. "You absolutely have to work with people from all kinds of disciplines and totally different disciplines, and it's kind of imperative."

"Most of the time, though, the first wave of technology doesn't get to artists and those pursuing social justice," said Gerry Flahive, a former NFB producer. "I hope, in this sort of revolution, people who want to be involved in co-creation of socially engaged documentary could get just as ready access to those kinds of tools that are, obviously, already being used for tracking us, for understanding how we behave, by corporations and governments. . . . You can swim in waves of data and patterns that might emerge and be playful with it."

New technologies can also be inaccessible to participants or potential audiences and may actually get in the way of production. When projects get overrun by technology, they run the risk of becoming solutions looking for problems. In her 2017 book *Twitter and Tear Gas*, Zeynep Tufekci documented how activists around the world have effectively used networked tools such as text messaging, Twitter, and

A gathering of Indigenous women in Peru, as part of the Quipu Project *(2015). Used with permission from María Court and Rosemarie Lerner.*

Google spreadsheets to strategically organize movements and tactically save lives. However, she wrote, "I have also seen movement after movement falter because of a lack of organizational depth and experience, of tools or culture for collective decision-making, and strategic, long-term action. Somewhat paradoxically the capabilities that fueled their organizing prowess sometimes also set the stage for what later tripped them up, especially when they were unable to engage in the tactical and decision-making maneuvers all movements must master to survive."[29]

Any and all projects working with technologies—co-creative or single authored—need to consider the platforms within which the media and relationships are embedded. Code and tech do not solve our preexisting social conditions. As Ruha Benjamin writes in *Race after Technology* (2019), "Zeros and ones, if we are not careful, could deepen the divides between haves and have-nots, between the deserving and the undeserving—rusty value judgments embedded in shiny new systems." But she warns that the notion of the "digital divide" also reproduces the cultural essentialist framework of inequity.

Participants from one of the Peruvian communities involved in the Quipu Project. *Used with permission from María Court and Rosemarie Lerner.*

Fireworks and cell phones light up the night at Tahrir Square as demonstrators celebrate the resignation of Egyptian president Hosni Mubarak on February 11, 2011. Photo by Jonathan Rashad. Distributed under Creative Commons Attribution 2.0 Generic License.

"A focus on technophobia and technological illiteracy downplays the structural barriers to access and also ignores the many forms of tech engagement and innovation that people of color engage in."[30]

Many co-creative teams we interviewed argued for site- and culture-appropriate technologies for the making of decentralized, organizing work. In the *Quipu Project*, teams used analog phone systems to organize and share anonymous stories among Indigenous communities in remote parts of Peru. Vojo, the phone system developed in part at the MIT Center for Civic Media, also has been used to organize and share stories among immigrant laborers in California to minimize the risk of exposing identities. "It is a privilege of those of us in [high-income] countries [to] have endless access to electricity and to every kind of device imaginable," said media scholar Patricia Zimmermann. "I actually think we have to think more like Malcolm X about this and say 'By any technology necessary.'"

IN CONVERSATION: APPROPRIATE MODELS FOR CO-CREATION

by CARA MERTES

Excerpt from an interview with Cara Mertes, Ford Foundation project director for Moving Image Strategies, International Programs, and formerly of its JustFilms creative documentary portfolio.

CARA MERTES: Co-creation is a concept that covers a broad set of practices, which revolve around equity, dignity, and justice. Supporting co-creation research, projects, and initiatives in nonfiction has been a long-term investment of mine. It's part of an effort on my part to more clearly articulate what a social justice cinema practice looks like in practice. Such a practice suggests that social justice–centered values are embedded in the process of story creation itself. This, in turn, is ultimately about power in the story-making process: who holds it, how does it move through the production and distribution process, what is done with it, who benefits and who doesn't. These central questions have animated my work in public media and at *POV*, the Sundance Institute, and the Ford Foundation.

A social justice moving-image practice seeks to recast embedded power relationships. It means moving from an extractive model of story making to one that reflects a realignment of power and agency within the process of story making as well as the reception of the project. It means focusing not only on compelling and original storytelling but also on creating public value or common good through the process and the product—what I have called a leave-behind in the community. This is an especially resonant approach at a social justice foundation, and JustFilms has proven to be a fertile laboratory for innovation in funding co-creative work.

This approach has implications for many aspects of strategy design, funding decisions, and ultimately the goals of the portfolio. As an example, time becomes as valuable as money in co-creation. Funding individual careers and leadership to experiment and iterate over many years and multiple projects is critical, as with Thomas Allen Harris as he honed his approach for what has become *Family Pictures, USA*. Supporting individual leadership and creative growth opportunities is another tool and has proven to be enormously impactful through the JustFilms Rockwood Fellowship. Many of these fellows, like Michael Premo, Jennifer MacArthur, Loira Limbal, and Opeyemi Olukemi, were nominated because they work across community strengthening and media making and rarely if ever receive recognition of their individual practices and leadership capacity.

All of these funding strategies speak to support for ideas, individuals, and entities that then apply themselves to co-creative practices. Finally, co-creation practices train practitioners in more sophisticated approaches to building truly representative and equitable communities, and at a time when democracy is showing its fragility under pressure from plutocrats and autocrats around the world, it is a powerful method of building power from the ground up by establishing a right to one's narrative as a widely held value.

Co-creation is an aspirational model. In my experience, it is difficult to achieve pure co-creation in narrative making, as narrative making itself involves a power dynamic rooted in trust, sharing, and ultimately a loss of control of the material on the part of the story-sources, as the final decision-making on the shape of the narrative is entrusted to the director/s.

Co-creation offers a way to interrogate traditional modes of representation, codified hierarchies, and their associated value chains. Because achieving a purely co-creative approach is difficult, JustFilms has supported projects and people who are exploring co-creative practices, but I would argue that it is extremely difficult to make and distribute work in the US that successfully disrupts the Western model of storytelling from inception to distribution. However, supported projects which illustrate how co-creative approaches can alter the design and impact of a project or initiative include major initiatives like the Detroit Narrative Agency, Electric South, and *Family Pictures, USA*. Film projects include *Decade of Fire* and *The Feeling of Being Watched*. Each employs co-creative practices which define the design and reception of the project.

Evaluative approaches of cultural strategies like co-creation have lagged behind evaluation in other sectors, I believe because they involve complex social processes, not linear or objective data. Social change is dynamic and unpredictable. Creative processes are also always human processes, and at their core they are based in emotion, not reason. One of the primary reasons for Ford supporting the *Collective Wisdom* field study was to begin to address this gap in knowledge and start to sketch a framework for better understanding what co-creation has been and could be and thereby strengthen its standing as an opportunity more funders feel comfortable embracing. Currently this lack of ability to fully predict, quantify, and classify the process and impact of co-creative approaches has marginalized the strategy. Rather than continuing down a purely numbers and scale-driven path to understanding impact, a new framework is needed which assesses rigorous data but equally values the way people and communities function—through trust, familiarity, respect, and ultimately alignment in basic values.

LESSON 8: TECH TRAINING, MEDIA LITERACY, AND CIVIC ENGAGEMENT

Providing community access to technologies—and even writing code—is core to many co-creative methodologies. Many twentieth-century co-creation projects handed over radio stations, video cameras, and editing systems to communities. When the World Wide Web arrived, digital redlining came with it; coding workshops and other ways of providing access to software and hardware popped up as a means of confronting these systemic barriers. Now with emerging and immersive tech, even greater challenges present themselves.

Salome Asega of the Iyapo Repository, a group that has media literacy and community training close to its core, noted that the learning curve for emerging tech is steep. "You can't just ask people to jump into making unless you're ready to provide a 101, a training of some sort."

Iyapo Repository workshops have gone through many iterations, from an idea-generating workshop to more hands-on opportunities with digital tools. The repository now offers a physical-computing track with Arduino, an open-source prototyping (testing) platform, and a VR drawing track with Vive and Tilt Brush so that participants can draw artifacts in 3D space. There is also a digital-fabrication track that includes 3D printing or laser cutting. "I can't make things with people using new technologies unless they also understand what's happening," Asega said. "I have to make it feel like it's not magic."

Increasingly, artists working with tech are writing their own code and making it available through open-source tools. "Tool-building is both an artistic pursuit and a

Community participants in a training session at NOVAC (New Orleans Video Access Center). Used with permission from Darcy M. McKinnon.

Working with VR at the Iyapo Repository (2017). Used with permission from Salome Asega.

vital practice for this field. It expands access to software and computational think-ing, and it supports the creativity of other artists through an open-source ethos," as reported in the 2021 National Endowment for the Arts study on technology as art.[31]

Some artists in this field split their own artistic and community practices and consider them two different modes of expression. But they see the community-facing work as fundamental to their roles. Jason Lewis and Skawennati, for example, run their own studio practice, producing Indigenous machinimas in the Aboriginal Cyberspace in Territories network. Concurrently they run SKINS, youth workshops in gaming, coding, and digital story creation in the Indigenous communities in Kahnawake (on the south shore of the St. Lawrence River in Quebec) and in Hawaii.

One arm of Tahir Hemphill's multifunctional studio in New York City involves a monumental archive of rap music along with other research and art involving AI, software, and social justice. But Hemphill is also committed to a strong community-engagement practice including, according to his website, "public-programming methodologies that invite participants to engage critically with cultural research, data collection, data ethics, implicit bias, data bias, and storytelling—elements that ultimately empower the critical thinking and decision-making that leads one to become more capable of establishing informed positions."[32]

A scene from Aboriginal Territories in Cyberspace, co-founded by Jason Edward Lewis and Skawen-nati. Used with permission from Jason Lewis and Skawennati.

However, tech training and media literacy may not be enough, argues Ronald J. Deibert. "Russian trolls, murderous Islamic extremists like ISIS, and far-right conspiracy theorists, like followers of QAnon, are all among the most 'media literate' entities of our age."[33] He insists that a focus in public education on "civic virtue," as Plato and Aristotle and Seneca debated—one's duty to the collective as a citizen—deserves as much attention in curricula as STEM but is almost entirely absent from our classrooms.[34]

IN CONVERSATION: ORIGINS OF SCRIBE VIDEO CENTER

by LOUIS MASSIAH

The Scribe Video Center in Philadelphia is one of the longest-running independent video centers in the United States. The center began workshop programs in 1982 and pioneered co-creative programs through training and literacy with the emerging technology of the time, analog u-matic video (a now obsolete video cassette format). Recently, forty-four short videos from the Scribe Video Center project were selected to represent the United States at the new

Filming for the Scribe Video Center which provides filmmaker workshops, a free documentary program for community groups, and youth programs to promote video and film as tools for community support and self-expression. Used with permission from Louis J. Messiah, Scribe Video Center.

Museum of Black Civilizations in Dakar, Senegal. We asked Scribe director Louis Massiah about the inspirations for this world-renowned center.

LOUIS MASSIAH: Oftentimes, a confluence of impulses and inspirations may get things started, but it is the reality of material circumstance that has a huge impact on shaping things. In the 1980s, there were a number of organizations, institutions, nationally, where people could come together and learn from each other and have access to equipment. We're talking about analog video and 16mm film, but we're also talking about technologies that were quite expensive and not that accessible. There weren't a lot of cameras and editing equipment floating around. And there certainly weren't a lot of places where the expertise and the knowledge of the craft of filmmaking could be shared—and, as importantly, locations that supported the camaraderie of media creation. Aside from learning the craft, cameras and production equipment were heavy, so you needed help carting equipment around.

I was in a graduate program at MIT beginning in 1979. So, it was immediately after leaving MIT in 1982 that I started Scribe. At MIT I was in the Film/Video section, which was largely a purist documentary film program, that is, 16mm film (photo-chemical) production from the cameras to flatbed editing, which was led by Ed Pincus and Ricky Leacock, although there were also folks in the program that were knowledgeable about the emerging video (photo-magnetic) technology, in particular Benjamin Bergery.

Prior to coming to MIT, I had worked at WNET in New York. While I was there, there was a subentity of Channel 13 called the Television Lab, with guest artists like Nam June Paik, Phillip and Gunilla Mallory Jones, Joan Jonas, Bill Viola, Alan and Susan Raymond, and Shirley Clarke. The roster of video makers included some of the most noted artists of that time. David Loxston and Carol Brandenburg were running the TV Lab, and John Godfrey was the technology guru. I have very clear memories of watching Paik edit at the TV Lab's 46th Street studio, when he'd just gotten back from Guadalcanal. Also, around that same time I would go to paradigm-shifting video screenings at the Kitchen and the Museum of Modern Art [MoMA]. Deirdre Boyle had programmed a pretty extraordinary series of video exhibitions at MoMA.

I was in my early twenties and was learning about video as an art form and also appreciating the particular aesthetics that were involved with video. Video to me seemed closer to the capturing of truth and reality than 16mm film. So, I was embracing video.

My initial reason to go to the MIT Film/Video section was on a quest to develop a visual language to look at ideas in science. I was thinking "Okay, MIT should be the place to do this," but when I arrived, I realized that an exploration of video linguistics wasn't part of the intellectual flow nor the interest of the Film/Video section, which was almost exclusively focused on cinéma vérité documentary storytelling and also 16mm film.

But I still wanted to work with video. I found this place in Boston at that time called BFVA, Boston Film Video Association, which allowed members to borrow equipment and shoot. So, the works I did at MIT were video works. And Benjamin Bergery was teaching this one course on video technology (waveform monitors, vectorscopes, etc.), which I found absolutely wonderful. His approach was about developing the art through really understanding the technology.

But media arts centers were actually few and far between in the US; Philadelphia did not have a media arts center. There was no BFVA there. Also, I had partially edited my thesis film in New York City at another media arts center, Young Filmmakers on Rivington Street. They had three-quarter–inch u-matic video-editing systems, which you could rent for an hourly fee. So, I edited my thesis between Young Filmmakers and in Boston, going back and forth. When I moved back to Philadelphia, I realized we didn't have the kind of center where video makers could come together or where people could have equipment access. That was one of the germs to the idea: "Okay, well, why not?"

It began by just knocking on doors and introducing myself to folks in film collectives in Philadelphia. There was a really wonderful exhibition program—that still exists—at International House; it's now called the Lightbox Film Center. It was programmed by Linda Blackaby. There was a collective called New Liberty with a very generous filmmaker, Lamar Williams. Also, there was a very influential film programmer, Oliver Franklin, who was extremely helpful to me in moving back to Philadelphia. He connected me to the Pennsylvania Council on the Arts, which was trying to see how a media arts center could be established in Philadelphia.

Scribe started very modestly. My day job was working as a staff producer at the public TV station in Philadelphia, mainly directing documentaries for local broadcast. Nights and evenings I'd offer workshops in video technology and production. I borrowed the Xeroxed handouts that Benjamin had put together. The people that came to Scribe were oftentimes artists in other media. The workshops were either free or low cost. We were given free space at Brandywine Workshop, a fine arts print atelier on Kater Street.

Community members gather around a computer for the Documentary History Project for Youth, Scribe Video Center, 2000. Used with permission from Louis J. Messiah, Scribe Video Center.

We were realizing that although we were definitely ethnically, racially, gender-wise diverse, the folks using Scribe tended to be people in the know. We [began to wonder], how do we enlarge the circle to people who don't necessarily give themselves permission to create? Often, for working-class people, if there is going to be any cultural offering or opportunity for cultural expression, you think about it for the children. Oftentimes adults don't give themselves permission to engage in creative processes.

So, that became something that we really wanted to address and to really think about. I should say, there are a number of people that were part of the early Scribe: Sandy Clark, Emi Tonooka, Joan Huckstep, Gary Smalls, Lamar Williams, and Carlton Jones. Toni Cade Bambara, the writer and filmmaker, had a huge impact on Scribe. She first came to take a workshop in 1986 and then ended up teaching a documentary planning and script writing workshop for nine or ten years there.

Now, in the mid-'80s, there was a housing and homelessness crisis. A lot of housing advocacy groups, including the National Union of the Homeless, would say "Hey look, can you help us document this action or this event?" We realized we were not a news service, we're not a production company, and we don't have the capacity to do all of this community production in service to these groups, but what we can do is we can teach people.

From the beginning, a central idea of Scribe was sharing knowledge. It's in the model of Prometheus. If one has had the opportunity to have knowledge—and I felt that the ability to be at MIT was a chance at knowledge—isn't it wonderful to be able to share it broadly with the community? And so we then began to develop these community media programs. The first one was a project called Community Visions, where we would put a community group with two filmmakers for a year. The filmmakers would be facilitators. They were not the producers; they were not the directors. They were instructors. The editorial control rested with the community group, but the facilitators would teach this community group planning, camera operation, lighting, sound recording, and editing.

Over time, we began offering multiple workshops in various aspects of time-based media production, offered to both emerging video makers as well as to working artists as well as a range of community media projects. Out of the Community Visions project, we developed a methodology to undertake a community history project.

Recently, we've been collaborating with a variety of Muslim educational, religious, and social organizations to create a history of Muslim communities in Philadelphia. And then in 2016 we completed a large project on the Great Migration that looked at community organizations created during the First World War, when large numbers of African Americans came from southern states to Philadelphia to work largely in the industries and to escape the violence and repression of the South.

Necessity is definitely the mother of invention. Producing a film is creative work. And producing a methodology is creative work.

I think it's still true today that a lot of folks involved in independent media are progressive and are involved in community organizing work. If you look at Visual Communications, Appalshop, or Third World Newsreel, or the Newsreel Movement more broadly from the '60s, or if you look at places like Downtown Community Television [DCTV], there's a commitment to use video and film as a way of working toward something better, to strengthen communities. Speaking of DCTV, I very much remember meeting Keiko Tsuno when she presented at MIT when I was still a student. Keiko and Jon Alpert started DCTV. I knew their work from my time at WNET. She came to MIT, and it had an impact on me. I definitely remember her presence and her presentation—and the conversation with her.

Also, the projects of Jean Rouch in West Africa were influential. He also came to MIT as a guest of Ricky Leacock. Although I have my own hard critique of some of the ways that he worked collectively with communities, he presented a model and a history of participatory filmmaking. I also was aware of the important work of the

Filming for the Precious Places *project in 2007, New Africa Center, a project at the Scribe Video Center. Used with permission from Louis J. Messiah, Scribe Video Center.*

Film and Photo League here in the US in the '30s. It's not that I was trying to necessarily model Scribe after the Film and Photo League or DCTV or Jean Rouch, but I know there is something to be learned from the history.

Our participatory design and methodology seemed like the work we needed to do and the way we needed to work.

LESSON 9: NEW MODELS OF FUNDING, EVALUATION, IMPACT, AND OWNERSHIP

Many of our interviewees admitted that they were limited to practicing co-creation within the margins of their budgets, work plans, and schedules. It is often invisible work, not rewarded, acknowledged, or accounted for financially, philosophically, or materially. Co-creation is overlooked in most of the current systems supporting documentary, the arts, and journalism in the United States and elsewhere today—systems that are geared to recognize and reward the work of individual authors and creators. Co-creation requires new progressive models of funding, evaluation, impact, and ownership.

FUNDING

"The funding in this country, and I speak as a funder," said Tabitha Jackson of the Sundance Festival (formerly of the Sundance Institute), "does not tend to fund the creative process. It funds the production. So, there's this huge bit in the beginning, which is the conception and the thinking, the language, the stimulus, and the space and the conditions for creativity, which we don't go anywhere near."

However, documentary scholar Mandy Rose, "some of the most productive outcomes emerge from projects in which there has been significant investment before the outcome was known." She cited as examples Video Nation at the BBC, the NFB's HIGH-RISE, and the *Quipu Project*, which was devised within the REACT Sandbox in Bristol. The Sandbox, funded by the UK Research Council, provided a context in which the Chaka Films team could think deeply about how creative technology might support the struggle of the Indigenous Peruvians who had been forcibly sterilized in the 1990s.

"[Development] is the last thing that many funders want to know about, but I think it's also really, really important," said Rose. "Fund right at the development

Participants collaborate during Sandbox Lab. Photo by Shamil Ahmed. Used with permission of Mandy Rose.

stage. It allows the ethical relationships of the project to be really thought through, mapped out, and defined."

Lisa Parks, a digital media scholar, said that she sees the same issue in academic funding. "Institutions and funders always say 'We want interdisciplinary collaboration.' Yet they don't really reward people for reflecting on that process itself. It's always the research or the deliverables or the artwork that's the most valued rather than trying to really understand how people work together in really interesting ways."

Loira Limbal of Firelight Media said that funding bodies should stop rewarding extractive approaches. "It's protected by the silence that it's shrouded in, and so I think if we could talk about it more, and if we could figure out ways where we could share . . . because I know there are a lot of filmmakers that are not working in that extractive way, that are working in a way that's more values-aligned, you know? But they're doing that, and they're hiding. . . . They're hiding it in their budget lines, or they're kind of finagling."

Limbal also suggested that agencies look beyond who is in front of and behind the camera and strive to develop an inclusivity plan that extends to the funding agencies themselves.

DNA's ill Weaver reminded us that it is important to ensure that new funding models for co-creative processes actually go to communities and people already co-creating rather than simply outlining new criteria for people who already receive funding. Long-term co-creative practices therefore require a radical reconsideration of how funding models can help to support and nurture rather than limit or contain the process.

EVALUATION

Where there is funding there is evaluation, usually in the form of reports and official reviews. But the prospect of evaluation can serve to discipline, limit, and normalize the work in advance in ways that ill serve many co-creative projects as well as marginalized communities.

Documentarian and scholar Elizabeth Miller proposed that the evaluation of co-creative projects needs to come from barometers set within the project itself. She said that the goal should be to

use evaluation as a process to deeply inform our next steps, [one] that is based on values and that takes into consideration relationships that were formed, not simply qualifying criteria like "How many videos did you create? How many people did you reach?" Rather, maybe

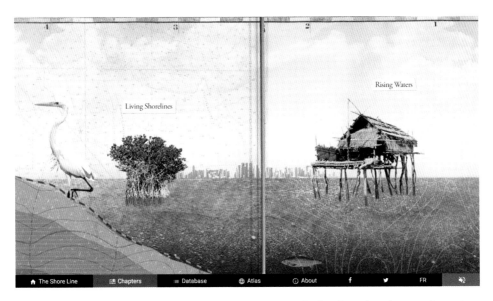

A still from The Shore Line *(2107), a web documentary by Elizabeth Miller, who advocates for evaluating works using barometers set within the project. Used with permission from Elizabeth Miller.*

it could be "What is the potential for the next stage of a collaboration? Where do we see this going? How can we go deeper?" I think we get caught into numbers, and I absolutely understand why we do, but we tend to think of impact in terms of . . . big audiences rather than deep audiences.

Along with the criteria, the very form of an evaluative report can be forbidding. Not all communities can necessarily write in the ways that funding reporting demands. The evaluation process of co-creation might better be served by other forms, such as oral testimonies, community meetings, and interviews. Further, the conventional definitions, impact models, and quantifications of success need fundamental reconsideration in co-creative works.

Authorship is also at issue in this process. Many co-creative teams complained that funders, festivals, and organizations fail to recognize collective authorship. "What if the 'you' is five people?" asked Hank Thomas Willis. "I think it really, often, winds up falling into the Rolling Stones or the Beatles [model], where one or two people get all the fame or credit, but the other people wind up filling [in] their background, so . . . they don't benefit as much, because maybe they're not as charismatic, or maybe they don't have the star power. I think the ethos can be harmed, which can be insidious to the overall headspace of a project."

This means change within institutions is needed as well. According to Monique Simard, a veteran commissioner and program director at many Canadian and Quebec cultural organizations,

You have to train people. For example, if you take film analysis, they will essentially read the script, see who's directing it, who are the lead roles, and that's it. It's very hard for them to conceive another way that a work will be eventually produced, directed, and produced. You have to change their mindset, so you have to train them. You have to bring them to think in nonlinear, nontraditional ways, and that means that they have to learn to evaluate these new things. . . . You just can't throw everybody out, and institutions are the ones that have to decide who will be selected. So, you have to bring in new people that will bring that experience but also train people to think differently, which is not an easy thing.

Co-creative evaluation requires building new evaluation and process maps from the inside out—within projects, within teams, and within the institutions and organizations that will support them.

IMPACT

Impact, engagement, and distribution strategies shift over the course of co-creative media making. One observer suggested that co-creation might seem like the inverse of conventional "impact" producing. However, Jennifer MacArthur, a documentarian and founder of the Impact Producers Group, proposed instead that co-creation can be part of the larger life cycle of a project: "Impact producing should keep frontline communities, social movements, and film 'subjects' front and center in the process."

Loira Limbal reported that Firelight Media is now convening large conversations with movement leaders early in the process rather than with a completed project:

We've been working with these folks, but usually we call them, you know, maybe at the rough-cut or at the final-cut stage. In this moment, part of what we've been thinking and what we've realized is that as horrible as everything that is happening in the United States currently is, it is also an opportunity to do solidarity-building work. . . . This moment is an opportunity to help create a vision that's more inclusive and where we understand that our futures are more connected than the current media present. We think that film and documentary film, nonfiction, can do that work. But that's not work that we want to do on our own. That's work that we want to be doing with the organizations that are on the front line, serving different communities and fighting the battles of the day.

Because co-creative work is iterative and often long term, impact, distribution, and feedback all coexist in a loop. Hank Thomas Willis recalled how burned out the *Question Bridge* team was by the time they put the project online. But going live online does not equate to the premiere or opening night for a film. As Willis commented, "The day you release a project online, it's only just beginning."

In co-creation, at least for many team members, distribution and impact will begin in a sense alongside the conception and launch of a process. Importantly, they cannot take over the creative process but must work symbiotically with each other.

OWNERSHIP

Challenges around ownership permeate the co-creative process. Elsewhere in this book, we stressed the dominance of the single-author model of media making and the importance of clearly defined agreements that help mitigate against extractive practices in which one creator or community's participation turns into someone else's property. Angie Kim sees ownership as a collective vision. "Our north star for systems change is ownership. The combination of owning real assets and having governance authority shifts power. Making online creative-content producers owners of their businesses—establishing what has become known as a platform cooperative—enables them to retain control of income-producing work and influence the trajectory of everyone's digital future."[35]

Intellectual property is a crucial issue. It includes alternative models to private ownership, such as open-source and sharing cultures (e.g., Copyleft free-use licensing and the software-creation platform GitHub). But these contemporary alternatives come with many precedents. Since the 1700s, cooperative ownership models have helped shape local economies across the world. From insurance schemes to agriculture to homes to credit unions and worker co-ops, collective ownership has enhanced many communities' well-being. Today, one in three Americans are members of some kind of co-op. In the art world, the West Baffin Co-op in Cape Dorset (on Baffin Island in Nunavut, Canada) has been a particularly successful model of co-ownership in Inuit art production and distribution since 1958. It is a remarkable north-south collaboration. In West Baffin, a coworking space provides stone cutting and printing studios as well as tools, training, and community space, while sales centers in the south offer broad distribution channels to the public. The West Baffin Co-op is an extension of a cooperative model that is commonplace in the Canadian north, including the joint ownership of local stores, tourist services, construction projects, and municipal-service providers.

Today the cooperative movement can take advantage of digital platforms as well, using technologies to manage and optimize some of the capabilities of co-ops. For

Artists Pudlo Pudlat and Kingmeata Etidlooie in front of the West Baffin Co-op in Cape Dorset, Baffin Island, Nunavut, in 1976. The co-op celebrated its sixtieth anniversary in 2019 as one of the oldest and most successful models of co-ownership in the art world. Used with permission from Tessa Macinstosh.

example, co-ops can agree on how long or how little a member may work on a given day and can be used to put caps within the system, automatically and transparently. Co-ops can also manage consensual decision-making practices. Platforms such as Loomio hold promise for simple, accessible virtual watering holes for large distributed decision-making and ownership. Loomio emerged from the Occupy Movement in 2012 and at first used Occupy hand signals as an interface. Since then, Loomio has turned into a social enterprise, building an online organizing tool for groups to discuss proposals by getting feedback from members on a specific proposition. Members can either agree, disagree, abstain, or block.

Ownership is on the brink of massive decentralization, according to the hype swirling around blockchain, cryptocurrencies, decentralized autonomous organizations (DAOs), and nonfungible tokens (NFTs). But are they solutions or snake oil? A system built on blockchain, like one built on the internet in general, can support any political agenda. There is potential, though, for "fractional, progressive ownership & collective production of art and livelihoods" and for "sharing and artwork [to] almost merge," according to the introduction to the 2018 collection of essays *Artists Re:Thinking the Blockchain*.[36] It's yet to be demonstrated how safeguards can be built into blockchain to prevent it from falling into familiar pitfalls of inequity, exploitation and bypassing public regulation. Still, many new forms of organization, ownership, and contracts could become available in custom-built sizes and shapes for fairer and more equitable and transparent organizational platforms, horizontally and nodally.

LESSON 10: ITERATION IS THE OUTCOME

"We'll fix it in post" is a joke and a reality in the media industry, a sector defined by tight production schedules, with projects moving from one stage to another in a linear fashion from development to preproduction to production and finally to "post." So, when something goes wrong on set, the only way to stay on schedule is indeed to move on and try to fix it later in the editing room. In co-creative projects, however, iteration—the process of going back to repeat stages—is common and expected. Making mistakes, experimenting, and applying new learning to a project at all stages is part of the emergent process of co-creation.

In his book *Emergence* (2002), Steven Johnson describes the connected lives of ants, brains, cities, and software. All of these, he contends, work in complex, emergent systems of continuous feedback from the bottom up in a way that informs the entire collective. Even though we call certain ants the queen, Johnson points out, they are busy laying eggs, not giving commands. He describes worker ants as having highly adaptive, decentralized intelligence that allows hive mechanics to interact and change the collective flow without centralizing command structures.[37]

Collective intelligence has even been found in slime mold, noted the artist Agnieszka Kurant. "I think we're living in fascinating times where all these discoveries are being made because similar discoveries have been made about colonies of bacteria. Dispersed cognition is yet another kind of characteristic of intelligence. It doesn't have to be happening with just one single brain or nervous system. It can be distributed, dispersed within a colony, but also within society, which correlates interestingly with the whole idea of collective memory, collective trauma of societies."

In *Emergent Strategy* (2017), adrienne maree brown considers emergence as "the way small actions and connections create complex systems, patterns that become ecosystems and societies. Emergence is our inheritance as part of this universe; it is how we change."[38]

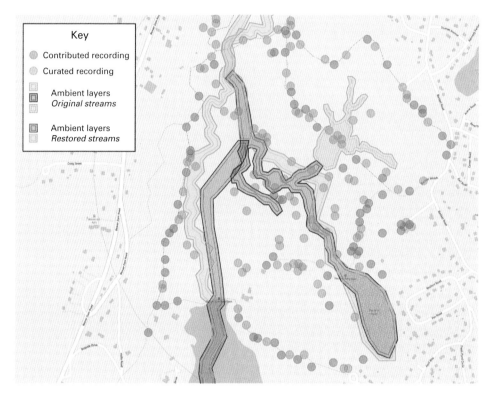

A map interface from Roundware, an open-source platform that allows users to collectively create continuous location-based audio experiences using mobile devices. Used with permission from Halsey Burgund.

Learning in the context of co-creative models is continuous and circular and embraces new information, new views, and new people entering the process at different stages. Much like the principles of Lifelong Kindergarten (a part of MIT's Media Lab), co-creation involves extending the play of the early stages of creative learning rather than rigidly formatting the process of learning into memorization, grades, protocols, and templates (see diagram of the Creative Learning Spiral).

While there is the potential for inefficiency and chaos in co-creation, so too is there great potential for iterative, emergent learning that builds on best practices rather than inflated, scaled-up models and generic tool kits.

Not all projects benefit from co-creation. Understanding what suits the specificity of a project and process is a matter of co-creators using listening skills, dialogue, and triage. When co-creation is appropriate, it is not leaderless; rather, co-creation

Creative Learning Spiral

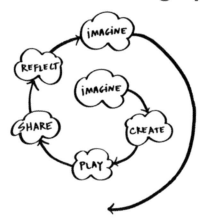

Mitch Resnick's diagram of the Creative Learning Spiral. Used with permission from Mitch Resnick.

requires a shift of roles and relationships that help to uplift and respect all the expertise in a group.

To avoid exploitative behavior, co-creative teams require time and strong, transparent decision-making mechanisms. They need mindful, meaningful, and often difficult conversations, negotiations, and feedback systems to ensure that the benefits of a project—financial, symbolic, or resource-based—are distributed equitably.

PHOTO ESSAY: *MIRROR | MASK*

Self-reflection is key to the iterative process of co-creation. Marissa Morán Jahn co-created *Mirror | Mask* as a series of photographs, videos, sculptures, and GIFs to explore how we see ourselves reflected and distorted in the Other. Jahn draws influence from Asian, African, and Greek rituals and dramaturgy as well as Western modes of self-presentation in an era of ubiquitous digital screens and mirrors. She notes that "as the phenomenologist Roger Caillois points out, masks are liberatory by the very fact that they conceal."

The Driver, *mask by Marissa Morán Jahn and Stephen Etende featuring Darlyne Komukama; collaborators include Alim Karmali, Moses Bwayo, and designer Gloria Wavamunno. Location: Nile River, Uganda. Photo by Marisa Morán Jahn, 2017. Used with permission from Marisa Morán Jahn.*

Jahn said that "the Driver wears a mask without eyes and features; you cannot locate her gaze. We imagined this character as Death incarnate, akin to Charon, the ferryman on the river Hades who transitions the living to the world of the dead. Perhaps she is part of your own psyche, always awaiting."

The Source, *mask by Marissa Morán Jahn and Stephen Etende featuring Aisha Abuova; photo assistance by Yerke Abuova and Abdel Mukhtarov. Photo by Marisa Morán Jahn, 2018. Used with permission from Marisa Morán Jahn.*

"The Source," Jahn stated, "was created with my intern Yerke Abuova, a high school junior originally from Kazakhstan. I accompanied Yerke, her twin Aisha, and her family on a trip to Central Asia, where we created a number of vignettes, learning the history of where we were through each other's eyes and through the camera's lens."

The Obstinate, *mask by Marissa Morán Jahn with Sydney Lee featuring Sultan Kusherbayev; photo assistance by Yerke Abuova. Photographed at Lake Kaindy, Kazakhstan, Jahn, 2018. Used with permission from Marisa Morán Jahn.*

"In Kazakhstan's Tian Shan mountains," said Jahn, "Yerke, Aisha and I went on a guided hike to this incredible lake created by an avalanche over a hundred years ago. The ice-cold, calcium-rich mountain water flooded the trees whose pine needles remain preserved underwater. Sultan Kusherbayev, the guide who showed us the way to this extraordinary place that he knows so well, wanted to pose as this creature. So, the photographs are a collaboration with place as much as with the people involved."

6

CONCLUSION: WHERE TO FROM HERE?

"Can't do it" or "been there, done that." For two decades in conversations with powers that be in the media industry, we encountered two standard responses to the prospect of co-creation. The first was "Oh, co-creation is impossible. Who would own it?" Intellectual property lawyers, journalists, producers, and decision makers dismissed the idea of collective creation as naive, impossible, and impractical. Authorship was foundational. Move on. The second type of response was "Oh, we do co-creation already." Media executives, funders, deal makers, and deal breakers who took pride in their latest innovation programs had convinced themselves that they had already solved co-creation (or tried it and failed), so it was time to move on. When we began the research, studies, and gatherings that led up to this book years ago, our aim was to challenge these dismissals, knowing that for anyone actually practicing co-creation, the hunger for relevant conversations is huge and continuous.

By the final phases of completing this book, however, the tenor of responses was markedly different. The world had changed, given the global health pandemic, the climate crisis, the massive movements to demand racial justice, and the threat of growing authoritarianism tied to surveillance capitalism. A perfect storm of accumulated wealth and power, competing expertise, and epistemological fragmentation has resulted in gridlock and system failure, exposing the fragility of the status quo. The urgency for new models has become undeniable.

Collage by Helios Design, with images from Marissa Jahn and Priya's Shakti Used with permission.

Precarious though this is, the situation has been exacerbated by the tapping and redirection of people's social impulses, just when they're needed the most. Social media platforms appear to scratch the itch of our mutual needs, offering just enough capacity for exchange, sharing, and interacting with our communities to siphon off the energy to do it in more meaningful ways. Through these networks, people's social behaviors are studied, gamed, exploited, and bent in ways that purport to be to their own benefit but clearly serve someone else altogether. These platforms are insidious because of not just their profit extraction and lack of transparency but also their displacement of meaningful collaboration and ritual.

And yet, rooted deep in human nature is hope as well as the capacity to act together. As noted at the start of this book, COVID-19 triggered perhaps the largest global act of co-creation in the history of humanity, as have social movements around the world that have risen in distinct and powerful choruses to call for justice. The enormity of the challenges on the horizon is such that only by pulling together will humanity survive, but in the process the world could come closer to new forms of liberation.

One such recent remarkable act of co-creation revealed a massive global series of spy operations, dubbed the Pegasus scandal. It started in 2016 when Citizen Lab, a human rights and tech research group, first identified spyware found on the phone of a human rights activist, Ahmed Mansoor. The group traced the spyware to Pegasus, developed by the Israeli cyberarms NSO Group. Mansoor likely installed the app on his phone unintentionally by downloading a link from a text; now it could read his messages, trace his phone calls, trigger his device's microphone and camera, and access his social networks. News organizations soon found that this technology was being employed by several governments, including the United Arab Emirates, Panama, and Saudi Arabia. But revealing Pegasus's true scope required a concerted co-created effort in 2021.

Armed with a data leak of 50,000 cell phone numbers, the Amnesty International Security Lab worked with the Citizen Lab, Forbidden Stories, and eighty journalists from seventeen media organizations in ten countries. Their coordinated findings revealed that Pegasus was being installed covertly on the phones of journalists, attorneys, human and environmental rights advocates, and even world leaders, resulting in direct reprisals, including murder, by some of the planet's most repressive political regimes. The networked character of the reporting thwarted attempts to suppress the coverage, because if one news source was censored, media outlets in other countries could keep publishing.

Meanwhile, Forensic Architecture, a research lab housed at Goldsmiths, University of London, and known for its interdisciplinary work with civic society groups as well as other artists, scholars, and more, used digital models and artificial intelligence to locate and synchronize source materials and narrativize them in legible and effective ways. Among other things, Forensic Architecture developed a navigable web-based digital platform, video investigations, interactive diagrams, a documentary (with Laura Poitras), and even an acoustic interface by Brian Eno to demonstrate the connections between Pegasus and real-world violence directed at lawyers, activists, and other civil society figures.

Pegasus represents the malicious side of the use of technology, but the Pegasus Project and Forensic Architecture show the power of co-creation and collaboration to push back and demand a semblance of justice.

With this book, we have tried to chronicle co-creation as it happens, drawing on diverse co-creative experiences and projects, learning from thought leaders, and making clear the terms, possibilities, and stakes. We focused on the practices of and the possibilities for co-creation across communities and an array of systems, but we also grounded our discussions in historical precedent. The point was to learn from the deep past of co-creation and to underscore its continuity with our present, of course, but also to reveal the historically specific nature of the regimes of attribution and control-through-ownership that dominate today's world. Single authorship and ownership are neither necessities nor givens; rather, they are for better or worse bound up in an economic order, itself part of a self-centered worldview. In this context, co-creation poses a paradox. It underlies the creation of our languages and great narratives yet cannot be found in our dictionaries. It acknowledges the power of the collective to sort through and address the complex problems of the day, yet it is too often drowned out by louder, more product-oriented voices. It is diverse by nature, drawing on multiple vantage points and values, yet is sold back to us in thin and piecemeal form, wrapped in the guise of expertise, authority, and intellectual property.

As we have noted throughout, authorship and attribution are not the enemy. They have important advantages, just as they have significant costs. They have also been written about extensively and codified into legal traditions and business models. And in a way, they pose a reverse problem to that of co-creation. Whereas co-creation is pervasive to the point of being invisible, ownership and authorship have been formalized as universalizing instruments that discipline and dominate our imagination.

We have endeavored to look beyond these constraints, finding inspiration in the work of communities, artists, scientists, and organizers who have charted a palette of robust alternatives. We have mapped a wide array of innovative precedents and projects. And we have done so cautiously, aware that like any instrument, co-creation is only as good as the purposes to which it is put. With so much demonstrated promise, then, why does co-creation seem so elusive? Why does it remain a verdant undergrowth in a habitat dominated by towering culture industries? We are left with the question of how co-creation itself might be better supported structurally and permanently.

WISE, COLLECTIVE, JUST

The many voices gathered together in this book are in harmony when they say that co-creation only becomes wise when it is bound to equity, justice, and respect. And while these terms are lofty going on abstract, makers and communities have found ways to concretize and operationalize them through techniques ranging from deep listening to community benefit agreements.

Wisdom, from this perspective, is necessarily collective and embedded in the larger social fabric. The disaggregation of decision-making from the social as well as the privileging of singular metrics, such as profitability and control, have led directly to some of the greatest threats to survival. What to do? "More of the same" is fast losing its value as an operating logic, and a little wisdom can go a long way in terms of helping to reframe priorities. The inclusivity of the social offers ways to build trust and solidarity. It is no wonder that methodologies related to co-creation are blossoming across disciplines, from architectural design to place making to the sciences. Duke Redbird's question "Is it wise?" needs to be bumped to the head of the queue.

One lesson looms large: complexity requires collectivity. Experts from any one domain inevitably risk replicating their own interests and paradigms, which is at best ineffective with a problem too vast to be pinned down, Gulliver-like, by the tenuous threads of disciplinarity. Collectivity is multiperspectival by nature and, when designed intentionally, offers a way to be inclusive. Divergent disciplines and methods provide a way to grasp complexity, but so too do the perspectives that emerge from different life experiences.

One of the great potential dividends of co-creation is trust. It requires shared frames of reference, transparency, and above all justice in order to thrive and is essential if we are to work with and depend on one another as a society. As we wrote this book, an environmental scan showed climate change deniers, antivaccination activists, QAnon conspiracy theorists, warmongers, and even flat earthers all pursuing media exposure and seeking their own agendas of hate and violence. In a perverse way, they too, in many cases, tap into a guise of co-creation to build trust and social bonds even if in opposition to the common good—a reminder that co-creation is not a panacea. The antidote to these niche appropriations may simply be a matter of scale, of a broader social embrace of the trust- and community-building work of co-creation. If journalists and those entrusted to create policies want their processes to be transparent and trusted, co-creation offers a way to suture epistemological divides, include diverse experiences, and build shared platforms that can help foster a renewed consensus of appreciation for verifiable facts and the necessity of ameliorative action.

Throughout the interviews, discussions, and gatherings we held for this project, we heard that co-creation insists on profound attention to process. The media industry has commodified the construction and telling of stories, so the prevailing standards of funding, researching, evaluating, and producing leave little room for process-oriented work. The current pipelines of production rely on and demand highly specialized, often extractive processes that are defined by deliverables and outcomes. Co-creation questions these modes of production and offers alternatives but ones that require time, trust, and iteration.

There are multiple dimensions within the ecosystems of co-creation, and stakeholders come from a wide variety of starting points. Co-creation is about the opportunities afforded by the meeting grounds, although boundary-setting by historically vulnerable communities often must be nonnegotiable. Co-creation can contribute to bridge building among divergent stakeholders as well as bonding within fixed groups. Tensions and distinctions are many. It is necessary to recognize that one person, one system, one discipline, or one agent (human or nonhuman) should or could not singularly hold the power to address some of the biggest challenges of this moment.

In this book, we have attempted to name and illustrate processes that are all too often invisible or implicit. At their best, these methodologies show that designs and solutions can be found in traditional and local knowledge as well as in movements, especially when thoughtfully intersected with the use of accountable research methods, open technologies, and appropriate engineering. We continue to gather concrete and detailed procedures and make them available to communities, festivals, organizations, funding bodies, and colleagues in other fields. We continue to listen and learn with and from the field, sharing what we know and refining our thinking about the operations of co-creation in various settings.

THE FINAL WORD: SUPPORT PROCESS

Overwhelmingly, we've found that the key to co-creation involves supporting and investing in process, not just in products. This imperative extends across individual projects, community initiatives, and institutional support and argues for systemic changes in the ways that media are produced and how they connect to social movements.

1. EXPANDING THE RESEARCH

There needs to be more research, from many sources, in order to map and understand the operations of co-creation in the context of our dominant culture, one predisposed to individual ownership, accumulation, and appropriation. We need to understand the implications of co-creation in a society of systemic inequity and in an era of fast-changing biological and technological developments. We need to continue to learn from historical and current human practices by studying and understanding business/organizational models, co-creation in diverse communities, co-creation around the world, ownership and intellectual property models, art collectives, cooperative economic models, transdisciplinarity models and matchmaking, art and artificial intelligence, deepfake and synthetic media, 3D modeling, and new nonhierarchical forms of convening.

2. A LIBRARY OF TOOL KITS AND CURRICULA

There is a need to create resources for teaching, sharing, and learning co-creative models. This process, which needs to avoid at all costs a "tick the box" approach, involves co-creative strategic planning that will build networks and hubs to document, organize, and compile an accessible library of existing tool kits (contracts, worksheets, community agreement forms), best practices, and modular curricula to record and share successes as well as failures. These networks and hubs should include media makers, community groups, nonprofits, private companies, public institutions, media institutions, and universities. These resources should be intended for professional development as well.

3. STRUCTURAL CHANGES AT INSTITUTIONS

Institutions, both public and private, must undertake more research and prototyping around co-creative efforts. The willingness to fail often is crucial. This will develop pathways for co-creative practices internally and methods to reach communities that already co-create. These processes must be ethical, just, transparent, and equitable. Co-creation is overlooked in most of the current systems supporting documentary, the arts, and journalism in the United States and elsewhere today, systems that are geared to recognize and reward the work of individual authors and creators. Co-creation requires new progressive models of funding, evaluation, impact, and ownership—you cannot expect to simply program co-creation within existing models.

4.   SPACES FOR INCUBATION AND PRODUCTION

More sustainable long-term programs, fellowships, workshops, streams, and incubators should be developed to facilitate co-creative projects that honor the processes, multiple partnerships, and time frames involved. The governance of these spaces and the projects needs to be interrogated. These sites need to provide adequate resources, mentors, cross-disciplinary supports, and witnesses, and intentional healing and trauma-informed practices should be implemented.

5.   NETWORKS FOR DISTRIBUTION

Spaces and networks for distributing co-creative projects need to be supported, including community centers, libraries, alternative spaces, schools, festivals, and universities as well as links among allied funders engaged in projects.

Together, we all share a vast history of co-creation. From early rock art to the development of sacred texts to the politicized twentieth-century newsreel collectives to the latest experiments in immersive technologies fueled by artificial intelligence, co-creation is remarkably commonplace. But it is also remarkably invisible. Before it is co-opted by digital empires and marketeers, there is still a chance to define it, claim it, and ground it to principles of equity, justice, and authentic collective models of ownership.

Media co-creation allows for new and better questions and for paths in which there are not always singular answers. Co-creation can enrich daily practice and demands self-reflection and forges harmonious, equitable relationships between partners within and across communities, beyond disciplines, and even with nonhuman systems, many of which no one fully understands yet.

Throughout the making of this study, primarily by carefully listening, we have been humbled and inspired by the phenomenal stories of co-creation and by the openness of all stakeholders to learn from each other and engage in courageous questioning. The conversations have been nuanced, messy, difficult, uncomfortable, exciting, and, above all, overflowing. Co-creation carries with it a profound respect for each person's unique expertise and also the knowledge that both the burden and the liberation of determining the future is collectively shared. No one person, organization, or discipline can determine all the answers alone.

Making can divide, alienate, and exploit but has the potential to be inclusive, equitable, and respectful. The latter conditions are far more conducive to the collective efforts it will take to address the immense challenges of structural inequality, exponential population growth, the Anthropocene, and the ever-diminishing resources that follow in their wake. In reaching beyond the mere sum of our collective intelligence, we stand a chance at finding our collective wisdom. Co-creation offers hope.

ACKNOWLEDGMENTS

This book is in itself a co-creation. It was shaped by many minds, hearts, and hands during the emergence of the Co-Creation Studio at the MIT Open Documentary Lab. The book is based on a field study that was originally intended as a small, short white paper, as suggested by Cara Mertes. It soon swelled into an enormous process of listening, research, writing, and rewriting with many voices and opinions to begin to give justice to the huge spectrum of co-creation. It's a project, we wish to acknowledge, that remains incomplete, and all errors and omissions are ours.

We are thankful to three editors and advisers without whom this book would simply not be possible. Jay Pitter first served as the guide for the 2018 Collective Wisdom Symposium connected to the study and then contributed significantly to the shape of the work on the written page and beyond. Carl Wilson and Dr. Suzanne Steele both offered their impeccable skills and talents as editors.

At every stage, we relied on the ideas and support of a dedicated team of research assistants from MIT's Comparative Media Studies graduate program: Deniz Tortum, Sue Ding, Sara Rafsky, Josefina Buschmann, Samuel Mendez, Kalila Shapiro, Beyza Boyacioglu, Srushti Santosh Kamat, and Andrea Shinyoung Kim. We are thankful to the 166 participants who took the time to grant us interviews and reviewed the draft field study.

The field study was also nourished through a relationship between the studio and the Detroit Narrative Agency. An enormous thanks goes to ill Weaver, paige watkins, and Juanita Anderson for their guidance and generosity in mapping and envisioning our work together, including the Community Benefit Agreement. We also express gratitude to Jenny Lee, Leila Abdelrazaq, and Sophia Softky of Allied Media Projects for all the support on the collaboration with the Detroit Narrative Agency.

Many in-person convenings along the way allowed for important group discussions on the themes in this book. We are thankful to the Detroit Narrative Agency for hosting the studio in Detroit two times, copresenting with us at the Allied Media Conference, and participating at the symposium at MIT. We also held conversations around our early findings at CPH:DOX, i-Docs, the Hot Docs International Film Festival, and the Skoll World Forum, and we are grateful to Professor Mandy Rose, Elizabeth Radshaw and Sandy Hertz for helping to organize these gatherings. We also had a chance to present some of our findings through live narrated presentations at MoMA Fortnight, thanks to Kathy Brew, and at the International Documentary Association's Getting Real Summit, thanks to Ranell Shubert. In 2018, we held the Collective Wisdom symposium at MIT, and we are grateful to Cheryl Gall for producing the very complex event.

For the section titled "In Conversation: Co-Creation and Equity, Five Media-Makers of Color Speak Out," the coauthors gathered online through a series of Zoom and collaborative writing sessions.

We brought the finished study into the world at three key gatherings, including the Climate Story Lab in New York City, thanks to Doc Society's Beadie Finzi and Jess Search; the Venice International Film Festival with Liz Rosenthal; and the Inaugural Future Imagination Summit at the Guild of Future Architects, thanks to Kamal Sinclair and Sharon Chang.

We also received invaluable feedback from all 166 interviewees, who were offered a chance to review the entire draft document. Additionally, the study was enhanced by the deep, long reads from and important conversations with Kamal Sinclair, Professor Elizabeth Miller, Professor Mandy Rose, Wendy Levy, Sam Gregory, Lina Srivastava, Jennifer MacArthur, Monique Simard, Gerry Flahive, Professor Patricia Aufderheide, Lisa Jackson, and Jessica Clark.

The visual component of the field study was significant, and we are thankful to Alex Whitholz and Heather Grieve at Helios Design Lab, as always, for the original illustrations and their translation into the cover and collages in the book. Co-creation came to life through an illustration by Michelle Hopgood, original photography by Kashira Dowridge as well as a huge amount of photographic research and archive clearance thanks to Samuel Mendez, Josefina Buschmann, and Srushti Santosh Kamat.

The MIT Press PubPub, an exciting new initiative to explore the future of reading, kindly took us on as one of its early projects to publish online. Catherine Ahearn and

Travis Cohen were remarkably supportive and enthusiastic, especially given the scale of visual materials we brought with us. MIT Press editors Doug Sery and Noah Springer kindly shepherded the project onto the in-real-life printing presses, with the support of acquisitions assistant Lillian Dunaj, production editor Marcy Ross, assistant production manager Kate Elwell, design manager Yasuyo Iguchi, publicist Zoë Kopp-Weber, along with the marketing expertise of Susan L. Clark, and publishing assistance from Cheryl Hirsch and Yvonne Ramsey at Westchester Publishing Services.

Once the field study was launched, we were so fortunate to continue publishing new related works at IMMERSE, an online publication dedicated to the creative discussion of emerging nonfiction storytelling. Jessica Clark, Carrie McLaren, Abby Sun and Ingrid Kopp gave us invaluable time, editing, energy, and laughter.

Our studio and lab exist within a remarkable community and lineage of thought and practice at CMS/W and MIT, including our colleagues, professors Sasha Costanza-Chock, Ethan Zuckerman, James Paradis, Fox Harrell, TL Taylor, Heather Hendershot, Eric Klopfer, Nick Montfort, Lisa Parks, Justin Reich, Paul Roquet, Paloma Duong, Ian Condry, and Ken Manning and the decades-long conversations with CMS/W cofounder and friend Professor Henry Jenkins. We deeply appreciate the new principal investigator of the lab and studio, Professor Vivek Bald. We thank the staff, including Sarah E. Smith, Jessica Tatlock, Gabrielle Horvath, Michael Gravito, and lab producer Claudia Romano for all the behind-the-scenes energy and work. We also thank Leila Wheatly Kinney for the inspiration and the first spark in the emergence of the Co-Creation Studio.

The Co-Creation Studio has been fortunate to be infused by the work and wisdom of fellows Assia Boundaui and Amelia Winger-Bearskin as well as key hands-on workshops with Violeta Ayala, Brett Gaylor, Sam Ford, Marissa Jahn, Lara Baladi, Cindy Bishop, Shirin Anlen, Dr. Julie Nagam, and Kerry Swanson of the Indigenous Screen Office, and Dr. Duke Redbird, all of whom have animated the questions central in this book. We also acknowledge the many fellows, too numerous to list, at the Open Documentary Lab throughout the years who have helped build our village.

This work is inspired and drawn from many years of experimentation at the National Film Board of Canada and the incredible co-creative team at HIGHRISE—Gerry Flahive, Heather Frise, Branden Bratuhin, Paramita Nath, Maria-Saroja Ponnumbalam, Sarah Arruda, Cass Gardiner, Deborah Cowen, Emily Paradis, Alexis Mitchell, Brett Story, Kate Vollum, Anita Lee, Silva Basmajian and David Oppenheim—as well as the residents of the high-rises at 2667 and 2677 Kipling Avenue in Toronto, Canada.

We are grateful to consider our funders as a great source of knowledge and wisdom. Our work is informed by the creativity and intellect of Kathy Im, Lauren Pabst, Cara Mertes, Chi-hui Yang, Jon-Sesrie Goff, Salome Asega, and Margaret Morton, and we honor the memory of our colleague Barbara Powell. We're thankful too for the support and thoughtful contributions of Tabitha Jackson, Richard Perez, and Sandy Herz at the Sundance Institute and the Skoll Foundation; Sheila Leddy and Diana Barrett at Fledgling, and Myriam Archard and Phoebe Greenberg of the PHI Centre. Dr. Charles Falzon and Dr. Charles Davies at Ryerson University supported the symposium and hosted Cizek as a distinguished visiting professor in Ryerson University's Faculty of Communication and Design during the research for the field study. This work has been made possible with the lab and studio's longstanding collaboration with the International Documentary Film Festival Amsterdam (IDFA) DocLab Research and Development Programme; special thanks to Caspar Sonnen, Yorinde Segal, and Najah Aouaki. And we celebrate the influence of George Stoney in his various guises as director of the National Film Board's Challenge for Change project, professor at New York University, and all-around wise mentor.

The field study's principal investigator is William Uricchio, coprincipal investigators are Katerina Cizek and Sarah Wolozin, and the field study was funded by the Ford Foundation's Just Films as well as the John D. and Catherine T. MacArthur Foundation, with contributions from Ryerson University, IMMERSE, the PHI Centre, the Fledgling Foundation, and MIT's Comparative Media Studies/Writing Department.

Finally, all the authors are thankful for the family, kin, and community with whom we co-create every day. You are (in addition to all mentioned above) the Bay Area Video Coalition, Charlene Bearskin, Ann Bennett, Black Public Media, Dorothea Braemer, Joe Brewster, the Center for Asian American Media, Corporation for Public Broadcasting, Claudine Brown, Kim Brundidge, Gillian Burnett, Bill Burton, Woo Jung Cho, Jiri Cizek, Ava Dixon, Sean Dixon, Leslie Fields-Cruz, David Haas, Ken Grossinger, Joan Huckstep, Rachel Kamel, Nehad Khader, Calvin A. Lindsay Jr., Cynthia Lopez, Bienvenida Matias, Vijay Mohan, the Mozilla Foundation, the National Endowment for the Arts, Boone Nguyen, Eamon O'Connor, Hébert Peck, Don Perry, Julian Ricco, Marja Roholl, the Seneca-Cayuga Nation of Oklahoma, Gary Smalls, Vincent Stehle, Margie Strosser, Third World Newsreel, Eddie Torres, Anna Van der Wee, Darren Walker, Noland Walker, Karen Warrington, Kathryn Washington, Deborah Willis, and Tristan Winger-Bearskin and the living legacies of Brian Winston, Jing Wang, Jacquie Jones, Emi Tonooka, and Toni Cade Bambara. It is in the daily acts of collective creation where we find the greatest sources of inspiration and humility.

APPENDIX: RELATED MEDIA PROJECTS

Title	Start Year	End Year	Type of Media	Description
#HACKED: Syria's Electronic Armies	2016	ongoing	game	An interactive journalism game in which users co-create by investigating cyberwarfare in the Syrian civil war while fending off hacks.
10 Minutes. 12 Gunfire Bursts. 30 Videos. Mapping the Las Vegas Massacre	2017		web-based investigative feature	A timeline and mapping of the October 1, 2017, Las Vegas mass shooting. This is a *New York Times* forensic visual investigation using thirty videos gathered from that night from a range of sources including cell phones and police body cameras.
18 Days in Egypt	2011	ongoing	web documentary	A crowdsourced documentary built on the interactive storytelling platform GroupStream. The documentary collects social media posts on the 2011 revolution and allows visitors to curate and publish their own media collections.
1979 Revolution	2015		game	A docugame about the Iranian Revolution, with the player embodied as a photojournalist in Tehran. In this docugame, users document the revolution and discover a world of chaos in which their own decisions matter.
99 Percent: The Occupy Wall Street Collaborative Film	2013		film	A collaborative documentary about the Occupy Wall Street movement. The documentary was co-created by one hundred filmmakers and artists throughout the United States.

(*continued*)

Title	Start Year	End Year	Type of Media	Description
A.A.I. (termites)	2018		AI	Agnieszka Kurant is a conceptual artist whose work blurs boundaries between fiction/reality and nature/culture. Her recent work employs Mechanical Turks to generate visual art as well as collaborations with nonhuman actors including parrots and termites.
Aboriginal Territories in Cyberspace	1997	ongoing	platform	A network of Indigenous academics, artists, and technologists who co-create new Indigenous-determined territories within the web pages, online games, and virtual environments of cyberspace. Efforts include a storytelling series, an ongoing games night, a modding workshop, machinima, and performance art.
Act of Killing, The	2012		film	A nonfiction film about mass killings in Indonesia in the 1960s. The film's subjects, perpetrators of the killings, designed and directed reenactments of the events.
And She Could Be Next	2020		film	Filmed by a team of women of color, this two-part documentary series follows the campaigns of women of color running for office in the United States in 2018.
AppRecuerdos	2017	ongoing	audio app	An audio-tour smartphone app that collectively gathers intimate stories about Santiago, Chile, during the 1970s and 1980s and places them in locales within the city. The app creates listening sites and connects public spaces with the memories encrypted into each locale.
AR Occupy Wall Street	2011		AR	An augmented reality protest that places layers of crowdsourced images, videos, and art over New York's financial district.
Are You Happy? Project, The	2010	ongoing	film	Inspired by *Chronicle of a Summer*, this project asks the same question in a contemporary global environment. Filmmakers around the world have been contributing to the project, all of them asking "Are you happy?"
Arqueología de la Ausencia	2014	ongoing	digital archive	This collaborative digital archive collects photographs, letters, testimonies, and other objects related to the detained and disappeared people from the time of the Augusto Pinochet dictatorship (1973–1990) in Chile.
Assent	2015		VR	Oscar Raby's VR piece about his father's experience in the Chilean army.

Title	Start Year	End Year	Type of Media	Description
BBC Video Nation	1993	2011	television series	A BBC television project of the Community Programmes Unit who distributed HI-8 cameras across the United Kingdom. Community participants kept the cameras for one year to film their everyday lives. More than 10,000 tapes were shot and sent into the BBC, from which 1,300 shorts were edited and shown on television.
Biidaaban: First Light	2018		VR	Rooted in the realm of Indigenous futurism, *Biidaaban: First Light* is Lisa Jackson's VR time-jump into a highly realistic—and radically different—Toronto of tomorrow.
Blackout	2017		VR	An ongoing participatory VR documentary in which New Yorkers share their stories in their own voices. *Blackout* aims to bridge social differences by illuminating moments in which lives intersect in the New York subway.
Blood of the Condor	1969		film	Grupo Ukamau's collective film traces the reaction of an indigenous community against a group of foreigners who, under the guise of development assistance, are forcibly sterilizing the peasant women.
Born into Brothels: Calcutta's Red Light Kids	2004		film	A documentary film about the children of Kolkata sex workers. Filmmaker Zana Briski taught the children photography and gave them cameras. The final film incorporates the children's photography and the classes.
Breathe	2020	ongoing	MR installation	Artist Diego Galafassi orchestrates co-creation on multiple levels. Using real-time biometric data from participants and environmental data from the exhibits' location, he creates visualizations of participants' breaths and encourages them to co-create the experience with each other.
BUFU: By Us For Us	2015	ongoing	installation	An artist and organizer collective of Black and East Asian queer, femme, and nonbinary people working to facilitate an international dialogue around the relationship between Black and Asian communities.
Build the Love You Deserve	2018		performance	Fei Liu's project is a performance series centered on the semifictional relationship between the artist and her DYI robotic boyfriend, Gabriel2052.

(continued)

Title	Start Year	End Year	Type of Media	Description
CareForce (formerly *League of Nannies*)	2012	ongoing	installation, performance, architecture	This project, co-created between artist Marissa Jahn and the National Domestic Workers Alliance, amplifies the voices of caregivers, America's fastest-growing workforce. *CareForce* consists of two mobile studios, an app, choreographed dances, printed works, and a cohousing project.
Casting the Vote	2020		documentary theater performance	A dinner, a party, a call to action, and a gathering, *Casting the Vote* is co-created by an investigative documentarian, a theatre director, and the audience to explore democracy and justice.
Challenge for Change	1967	1980	film and video series	A trailblazing participatory, social change–oriented documentary program, which consists of 145 films and videos in both French and English, developed by the National Film Board (NFB) of Canada. The filmmakers were among the first to exploit portable video technologies and challenged audiences, subjects, and filmmakers to confront a wide spectrum of issues, from poverty to sexism to marginalization.
Changing Same	2021	ongoing	VR	A VR piece co-created between a documentary film studio and a volumetric filmmaking team, including a vast array of creative technologists, about racial injustice in the United States.
Chinatown Art Brigade	2015	ongoing	installations	The *Chinatown Art Brigade* is a New York City collective of artists, media makers, and activists creating art and media to advance social justice, together with tenant unions and grassroots movements, fighting evictions, and displacement.
Chronique d'un été	1961		film	*Chronique d'un été* is a reflexive documentary film by anthropologist Jean Rouch and sociologist Edgar Morin wherein the subjects participated in the production of the film, interviewed others, and discussed and commented on the editing of the film.
Coded Bias	2020		film	*Coded Bias* is documentary that explores the fallout of MIT Media Lab researcher Joy Buolamwini's discovery that facial recognition does not "see" dark-skinned faces accurately and her journey to push for the first-ever legislation in the United States to govern against bias in the algorithms that impact everyone.

Title	Start Year	End Year	Type of Media	Description
Collisions	2015		VR	*Collisions* is a VR experience, centered on Indigenous elder Nyarri Morgan of the Martu tribe, offering viewers insight into his sighting of an atomic test in the 1950s in western Australia. The film is a message from Nyarri, in his own words and on his terms.
Conducttr	2010	ongoing	MR platform	A beyond-human, mixed-reality platform that is used to develop and deploy immersive stories and games across multiple channels and connect digital and physical worlds.
Conversations with Bina48	2014	ongoing	film series, performance	An ongoing series of filmed conversations between artist Stephanie Dinkins and one of the world's most advanced social robots.
Copwatch	2014	ongoing	platform	A community defense tactic for monitoring, documenting, and deterring abusive policing in different neighborhoods in New York City.
Coral Project, The	2014	ongoing	platform, tools	The Coral Project brings journalists and the communities they serve closer together through open-source tools and strategies.
Counted, The	2015	2016	platform	A crowdsourced journalism project to document cases of death at the hands of US law enforcement. *The Guardian* works with readers to gather verified data and report on these fatalities in response to a lack of comprehensive recordkeeping from the US government.
Crip Camp	2020		film	This co-created film follows the evolution of the disability rights movement in the United States, starting in 1971 at a summer camp for teens with disabilities—one of whom is the co-director of the film.
Crowd-Sourced Intelligence Agency	2015	ongoing	web documentary	An interactive artwork that allows users to perform the role of an intelligence analyst through an online interface. The goal of the project is to help people understand both the effectiveness of open-source intelligence processing and its impact on our privacy.
Curious City	2012	ongoing	platform	A collaborative news experiment at Chicago's public media station, WBEZ. Based on questions posed and voted on by listeners, the collaborative conducts research and creates journalistic stories.

(continued)

Title	Start Year	End Year	Type of Media	Description
D.O.U.G.: Drawing Operations Unit Generation	2015	ongoing	AI	Sougwen Chung's Drawing Operations Unity Generation (D.O.U.G.) is a study of collaboration between a human and a robotic arm machine.
Digital Violence	2021	ongoing	multi-platform	A presentation and analysis of data and testimonies by Forensic Architecture and partners to reveal a set of mass surveillance operations around the world using Pegasus spyware developed by the NSO Group in Israel.
Dimensions in Testimony	2014	ongoing	MR	A collection of testimonies designed to be experienced interactively. Viewers ask questions and receive answers from survivors of the Holocaust and other genocides who have been filmed to answer thousands of questions. A natural language processing system uses participant's questions to pull answers from the database co-created with the participants, AI, and the audience.
Dream	2021		digital, live performance	Theater, music, and emerging technology combine in the Royal Shakespeare Company's co-creative exploration into the future of live performance, where viewers follow a sprite on a moonlit journey through the night until dawn in a virtual forest. *Dream* is a demonstration project from five major partners under the Audience of the Future consortium.
Do Not Track	2015		web documentary	A personalized documentary web series on privacy and the web economy.
Edmond de Belamy	2018		AI	Art Collective's AI-generated artwork, the first of its kind to be auctioned at Christie's.
Electionland	2020		platform	A collaborative journalism project that covers voting access, cybersecurity, misinformation, and election integrity in the 2020 elections. Rather than covering the race and the results, Electionland's goal is to document, nationally and in real time, voting impediments and disruption.
Empire: An Exploded Feature Documentary	2014	2018	web documentary	Shot in ten countries over the course of four years, *Empire* is a webdoc that explores the impact of colonialism on today's world.
Everydays: The First 5,000 Days	2021		JPEG	Created over five thousand days by the artist known as Beeple, this collage was the first digital artwork ever auctioned at Christie's and sold for US$69 million.

Title	Start Year	End Year	Type of Media	Description
Eviction Lab	2017	ongoing	database platform	A lab at Princeton has built the first nationwide database of evictions open to the public. People can create custom maps, charts, and reports. The Eviction Lab is a team of researchers, students, and website architects who advocate for housing as a human right.
F for Fake	1973		film	Orson Welles's controversial fake documentary about fraud and fakery in which audiences discover that the narrator is fake at the end of the film.
Family Pictures USA	2009	ongoing	television series, performance, installations	Artist Thomas Allen Harris co-creates a living family picture album of America by traveling across the country and inviting community members to share images and stories from their personal family archives. The resulting work involves live interactive performances, documentary films, web projects, and a PBS TV special series.
Father's Lullaby, A	2019	ongoing	installations	A series of interactive public installations, community-engaged workshops, and a participatory website by Rashin Fahande to interrogate the structural violence of mass incarceration. Participatory, intimate interviews, songs, and lullabies offer poetic meditations.
Feeling of Being Watched, The	2018		film	When journalist Assia Boundaoui investigates rumors of surveillance in her Muslim American neighborhood in Chicago, she uncovers one of the largest FBI terrorist probes conducted before 9/11 and reveals its enduring impact on her community.
Filmmaker-in-Residence	2004	2008	web documentary, installations, performances, film	A five-year co-created program at the NFB led by Katerina Cizek at an inner-city hospital. The project partnered doctors, nurses, researchers, and patients with media makers and was inspired by the NFB's Challenge for Change and enabled by the web revolution.
Fireflies: A Brownsville Story	2020		VR	A documentary VR game developed by the Brownsville Community Justice Center and the Peoples Culture collective. Fireflies explores an ongoing rivalry between public housing developments. Players from both sides of the conflict work together to explore the histories and dreams of the community's residents.

(continued)

Title	Start Year	End Year	Type of Media	Description
Folk Memory	2010	ongoing	film, performance	An ongoing oral history and community engagement project to collect testimony from survivors of the Great Famine (1958–1961) in rural China. Wu Wenguang's studio in suburban Beijing, Caochangdi Station, is the home of this project. Young filmmakers have been to 246 villages in 20 provinces and interviewed more than 1,220 elderly villagers.
For Freedoms	2016	ongoing	installation	Art exhibitions, billboard campaigns, and public engagement programs across the United States, that leverage art as a means of creating conversation and encouraging analysis of politics.
Forensic Architecture	2011	ongoing	platform	A research agency that analyses violations of human rights and international humanitarian law to provide evidence for prosecution teams. The agency's interdisciplinary team of architects, scholars, artists, software developers, lawyers, and journalists work on a variety of projects including navigable 3D models of sites of conflict as well as reenactments and interactive maps.
Fort McMoney	2013		game	An interactive documentary game by David Dufresne that introduces the user to the myriad and competing perspectives of Fort McMurray, Alberta, Canada, the third-largest oil reserve in the world.
Gather	2017	ongoing	platform	An online platform that links together community-engaged journalists across the United States to learn from each other, collaborate, and make new connections to serve the mission of making journalism more responsive to diverse communities' needs.
Gee's Bend Quilts	1800s	ongoing	textile	The collective work of quilts of the women of Gee's Bend, a small Black community in Alabama bordered by a river on three sides. Creative works from several generations of families can be found today, and the quilts are notable for their bold visual style and masterful display of technique.
Ghana ThinkTank	2006	ongoing	installations	An international art collective that collects problems within communities in the United States and Europe and sends them to think tanks including a group of bike mechanics in Ghana, a rural radio station in El Salvador, Sudanese refugees seeking asylum in Israel, an artist collective in Iran, and a group of incarcerated women in the Boston penal system.

Title	Start Year	End Year	Type of Media	Description
Glass Room, The	2016		installation	A pop-up exhibition that explores data and privacy and has been developed by the Tactical Tech Collective. The experience can be set up in any space and by anyone and has been hosted in nearly one hundred libraries, schools, festivals, and organizations worldwide.
Granito: Every Memory Matters	2012		documentary, archive	An online intergenerational and interactive public archive of memories that attempts to uncover the history of the Guatemalan genocide.
Hearken	2015	ongoing	platform	A platform for news organizations to engage with their audiences over the entire lifespan of a story. Using Hearken, audiences can provide feedback and insight on in-progress stories.
Heirloom	2014	ongoing	bio-art	Heirloom grows living portraits of Gina Czanercki's daughters from cells that have been cultured from a single sample taken from their mouths in 2014. The cells are grown on delicate glass casts of their faces in a life-support system that provides the best conditions for growth outside the lab.
Hello Lamp Post	2013	ongoing	platform	A platform that invites people to strike up conversations with street furniture using text messages or Facebook Messenger, bringing to light the hidden stories of their cities. The team also develops urban co-design experiences.
Her	2013		film	Hollywood science-fiction film in which the protagonist falls in love with the (nonhuman) operating system in his computer.
Hidden Hunger	2014		multi-platform	Capital Public Radio (CapRadio) led community partners in a multiplatform and participatory media project to address food insecurity in Sacramento County.
HIGHRISE	2008	2015	web documentary series, films, installations, performances, audio	A seven-year multimedia, collaborative documentary exploring vertical living around the world. The documentary collection is led by Katerina Cizek at the NFB, co-created with high-rise residents, architects, urban planners, creative technologists, academics and media organizations such as wired.com and the *New York Times*.

(continued)

Title	Start Year	End Year	Type of Media	Description
Hollow	2013		web documentary	The creators worked closely with thirty residents in rural West Virginia to tell the story of the community. The result was an interactive web documentary as well as a community newspaper/blog.
Housing Problems	1935		film	*Housing Problems* is a short British propaganda film directed by Edgar Anstey and Arthur Elton and produced by "the father of documentary," John Grierson.
How a People Live	2013		film	Commissioned by the Gwa'sala-'Nakwaxda'xw First Nation, the one-hour documentary *How a People Live* traced the history of the Gwa'sala-'Nakwaxda'xw First Nation, forcibly relocated by the Canadian government in 1964 from their traditional territories on the coast of British Columbia.
Hyperbolic Crochet Coral Reef Project	2005	ongoing	installations	This project brings together community art making, science, and mathematics to create handicraft coral reefs around the world. It includes printed guides and workshops that allow people to understand the mathematical structure of a coral reef so they can create a crochet reef on their own.
IBM Watson Trailer	2016		film	IBM Watson, the AI developed by IBM, created a trailer for the horror film *Morgan* about the dangers of AI by drawing on trailers from one hundred horror movies, to identify visual, sound, and scene compositions that would fit the structure of a suspense or horror film.
In Event of Moon Disaster	2020		AI video	A video that deepfakes President Richard Nixon to deliver a speech he never actually gave. The film and its online packaging aim to educate the public about manipulated media.
Inverse Surveillance Project	2020	ongoing	AI, platform, installation	Assia Boundaoui's hybrid documentary, multimedia, AI project, co-created with her immigrant Muslim community in Chicago, that repurposes thousands of declassified FBI records collected during decades of surveillance as raw material for creating a counternarrative and reimagining the secret archive.

Title	Start Year	End Year	Type of Media	Description
Invisible Monument	2015		audio	*Invisible Monument* is a series of participatory audioscapes that record the world's social uprisings. As a part of Lara Baladi's ongoing project Vox Populi, recordings are geolocated in spaces across the world where specific social movements have risen. Listeners can experience and contribute to the project.
ISeeChange	2014	ongoing	platform	A participatory platform building community around documenting and understanding climate change, co-created with citizen scientists, journalists, community organizations, and NASA.
It (Still) Ain't Half Racist, Mum	2019		film	A project that puts the 1979 television essay *It Ain't Half Racist, Mum* (by Stuart Hall and Maggie Steed) in conversation with contemporary artwork and a panel discussion around race and representation in the media today.
It Gets Better	2010	ongoing	platform	An internet-based collection and campaign of testimonies by LGBTQ+ adults for youths.
Iyepo Repository	2016	ongoing	installations	A collection of both digital and physical artifacts that project the future of people of African descent. The repository offers an opportunity for people of African descent to reimagine race and identity through artifacts that will travel to the future.
Johnny Cash Project, The	2010		video	A crowdsourced collective portrait of Johnny Cash where visitors can contribute a frame to the music video for *Ain't No Grave*. The project randomly pulls frames submitted by users, creating a video that is unique each time it plays.
La Commune (Paris, 1871)	2000		film	A film by Peter Watkins that is a historical reenactment in documentary style and features a large mostly nonprofessional cast. Watkins had the cast conduct much of their own research for the project.
Land Art	2018		installation	This project was worked on by seven groups of artists and installed in the Iranian desert. Roughly two thousand people from different cities joined to reunite local people, artists and authorities with nature and environment.

(*continued*)

Title	Start Year	End Year	Type of Media	Description
LAUREN	2017		performance	*LAUREN*, is a performance piece created by Lauren McCarthy. *LAUREN* is a co-creation between a resident in their own home and a virtual assistant (like "Alexa") controlled by Lauren, the artist, who unleashed a collection of smart devices to watch over residents at all times.
Learning to See	2017	ongoing	AI	Developed by the artist Memo Akten, *Learning to See* is an ongoing series of works that use machine learning algorithms to co-create visual scapes as a means of reflecting on ourselves and how we make sense of the world.
Living Los Sures	2014	2019	web documentary	*Living Los Sures* was a multimedia documentary project that responded to the 1984 documentary *Los Sures*, a cinema vérité portrait of South Williamsburg, New York, and the largely Puerto Rican and Dominican communities that lived there. Produced over five years by sixty artists, the project includes a cinematic people's history in the form of a shot-by-shot analysis of the film by its subjects and local community members.
Localore: Finding America	2015	2016	multi-platform	Fifteen independent producers embedded at public radio and television stations across the United States launched Localore: Finding America in 2015. The project concluded in the summer of 2016, although many of its projects are ongoing.
Machine to Be Another, The	2014		VR	*The Machine to Be Another* offers users an opportunity to virtually experience other perspectives by giving access to the body and mind of another person.
Make the Breast Pump Not Suck	2018	ongoing	installation	A collaborative project between mothers and allies working to improve maternal health that includes hackathons, a policy summit, and research. The group aims to improve breastfeeding technology and support by representing women of lower socioeconomic backgrounds, women of color, and LGBTQ parents.
Man with a Movie Camera	1929		film	Soviet filmmaker Dziga Vertov's landmark experimental documentary intercuts Moscow, Kyiv, and Odesa, analyzing both urban life and the filmmaking process.

Title	Start Year	End Year	Type of Media	Description
Man with a Movie Camera: Global Remake	2007	ongoing	web documentary	Crowdsourced recreation of Dziga Vertov's celebrated film.
Mapping Main Street	2008		platform	A mapping project of all the Main Streets across America.
Mapping Memories	2007	2012	platform	Refugee Youth participatory media project in Montreal, Canada.
Marrow	2018	ongoing	AI	In a study of the concept of "family," machine learning models become anthropomorphized, with emotional rules generated from mathematical logic. shirin anlen's project incorporates a mix of research facilitating human-computer collaboration through text-to-image generators and a participatory art installation focused on stock images of families.
MIRROR | MASK	2018	ongoing	installations	A project that explores how we see ourselves reflected (and distorted) in the Other. Creating the masks and photographic tableaux with various collaborators, artist Marisa Morán Jahn draws on Asian, African, and Greek rituals and dramaturgy as well as Western modes of self-presentation in an era of ubiquitous digital screens and mirrors.
Monument Quilt Project	2013	2019	textile, installation	A public healing space by and for survivors of sexual assault and abuse. It is an ongoing collection of their stories. The stories are created on red fabric squares, which are added to a collective quilt and displayed in city centers across the United States as a demand for a new culture around sexual assault and abuse.
Motive.io	2007	ongoing	tool	A tool for developing location-based augmented reality apps and games.
Nanook of the North	1922		film	An early and influential narrative documentary film by Robert Flaherty set in Ungava Bay, now the subject of discussion around ethics, agency, ethnography, racism, and the decolonization of cinema.
Navajo Film Themselves	1966		film	A documentary film series resulting from a two-month workshop with Navajo students in Arizona. The films cover aspects of the students' local community, including weaving processes, a local lake, the regional landscape, a silversmith, and traditional medical practices.

(continued)

Title	Start Year	End Year	Type of Media	Description
NeuroSpeculative AfroFeminism	2017	ongoing	VR	A three-part digital narrative by the Hyphen-Labs team incorporating product design, virtual reality, and neuroscience. The products range from sunblock for traveling through the multiverse to earrings embedded with cameras that offer protection and visibility. The VR experience takes place in a neurocosmetology lab where Black women are pioneering neuroscience techniques.
Not the Only One	2021	ongoing	AI installation	Stephanie Dinkins has created an AI entity with tensor flow, an open-source software library for machine learning, and speech-recognition technologies. She trains it on data from three generations of her family archives to create installations and performances.
Omnia per Omnia (specific *D.O.U.G* project)	2018		AI performance	A co-created, improvised drawing performance between Sougwen Chung and a swarm of robots. The robots' movements are guided by the density, dwell, direction, and velocity of overall movement perceived in public camera footage of New York City streets.
One Day on Earth	2010	2012	film	*One Day on Earth* is a crowdsourced documentary showing moments in every country on Earth, all shot on a single day. The United Nations and over sixty nonprofit organizations participated in its making. The first film was produced in 2010, and subsequent films were produced in 2011 and 2012.
Ontroerend Goed	2007	ongoing	performance	These theater performances do not create alternate realities; rather, they take place in the present moment through a unique blend of experimentation with form and audience involvement.
Otherly	2021		film series on Instagram	A series of seven short documentaries about finding one's place in the twenty-first century. Using universal themes such as love, inclusion, and loss as entry points, seven female, nonbinary, and genderqueer creators have crafted films that are at once timeless and yet, by definition of their form, ephemeral.
Panama Papers	2016		platform	A global collective of investigative journalists from 107 global news organizations who joined forces in 2016 to interpret the largest data leak in history. The Panama Papers brought down governments and presidents and marked the biggest effort ever in journalism to collaborate rather than compete.

Title	Start Year	End Year	Type of Media	Description
Paper Tiger Television	1981	ongoing	TV station	A volunteer video collective. Paper Tiger Television carries community screenings, video production workshops, distribution of public access series, and advocacy work to challenge corporate control of media.
Paradise Papers	2017		platform	A major global collaboration reveals secrets from one of the world's most prestigious offshore law firms, a specialized trust company, and nineteen company registries in secret jurisdictions.
Pegasus Project	2021	ongoing	platform	A major global collaboration between human rights organizations and news media to uncover and analyze covert surveillance operations using Pegasus software.
Performing Statistics	2015	ongoing	installations	Connects incarcerated teens with artists, designers, educators, and Virginia's leading policy advocates in an effort to transform the juvenile justice system.
Prison Valley	2010		web documentary	An interactive web doc about a prison in which visitors can join live online debates and exchange emails with characters.
Prison X	2021		VR	VR experience that takes participants inside a magical realist storyworld, both mythical and profane, of a Bolivian prison. Co-created by documentarians, technologists, sound artists and a fashion designer.
Priya's Shakti	2014	ongoing	AR, comic book	An interactive digital comic book about a gang-rape survivor turned superhero. The project includes an exhibition of an augmented reality comic book incorporating real-life stories and voices of Indian sexual assault survivors.
Project Syria	2015		VR	Commissioned by the World Economic Forum, this VR project put the audience on scene, where they could experience the plight of Syrian children in the civil war.
Queering the Map	2017	ongoing	platform	A community-generated mapping project that documents queer memories and histories in relation to physical space.
Question Bridge	2012	ongoing	installations	A transmedia project bringing together Black men across the United States to exchange ideas and explore the meaning of Black identity in the country today.

(*continued*)

Title	Start Year	End Year	Type of Media	Description
Quipu Project	2015		web documentary	The co-creators worked with local community organizations and indigenous Peruvian women to document stories of forced sterilization using Vojo and community workshops to collect and record the stories.
Radio Diaries	1996		audio	Encourages citizen journalism by supporting people to tell their own stories, create sound portraits, and report on their own histories.
Rape in the Fields	2013		film	A collaborative transmedia documentary project involving a number of different organizations that expose the abuse of female agricultural workers.
Raqqa Is Being Silently Slaughtered	2014		platform	A citizen journalist group reporting Syrian war news and human rights abuses by Islamic State of Iraq and the Levant (ISIL) and other forces occupying the northern Syrian city of Raqqa, which ISIL used as its de facto capital. Raqqa Is Being Silently Slaughtered has described itself as an "nonpartisan and independent" news page.
Raqs Media Collective	1992	ongoing	installations	A New Delhi–based collective of artists who make contemporary art and films, curate exhibitions, edit books, stage events, and collaborate with architects, computer programmers, writers, and theater directors.
Report for America	2017		platform	Sends journalists to newsrooms across the country to report on local communities and underrepresented stories.
ROUND: Cambridge	2012	ongoing	audio installation	A sound art installation highlighting public art that allows visitors to leave geo-tagged commentary for other participants to discover through a specially designed app.
Roundware	2007	ongoing	platform	An open, contributory, location-aware audio framework created by Halsey Burgund.
SGaawaay K'uuna (Edge of the Knife)	2018		film	An Indigenous narrative film co-created by a film company, an Indigenous government and a university. It's the first feature film spoken solely in Haida dialects, which are endangered languages.
SOS Media Burundi	2014	ongoing	platform, audio	A collective of journalists that emerged spontaneously within forty-eight hours of the destruction of radio stations during the May 2014 coup attempt and continues to provide anonymous coverage of events in this troubled country.

Title	Start Year	End Year	Type of Media	Description
Sandy Storyline	2013		web documentary	A community-generated narrative of Hurricane Sandy that used new storytelling platforms, such as Vojo and Cowbird, to collect audio, video, photography, and textual stories from residents and citizen journalists.
Say Something Bunny!	2017		theater	A one-woman show by Alison S. M. Kobayashi, based on a found audio recording by a New York family in the 1950s, that interacts with the audience.
Seeing Is Believing: Handicams, Human Rights and the News	2002		film	A film co-created by Katerina Cizek and Peter Wintonick covering the impact of consumer video technology on human rights activist efforts around the globe.
Shoreline, The	2018		web documentary	A web doc by Elizabeth Miller exploring the environmental issues where the land meets the sea.
Stories of Our Lives	2013	2014	film	A Kenyan anthology comprising reenacted versions of true stories from people identifying as gay, lesbian, bisexual, transgender, and intersex in Kenya.
Syrian Archives	2014	ongoing	archive	A Syrian-led and -initiated collective of human rights activists dedicated to curating visual documentation relating to human rights violations and other crimes committed by all sides during the conflict in Syria.
Tatsuniya	2017	ongoing	film	A collaboration between artist Rahima Gambo and ten girls from the Shehu Sanda Kyarimi government school in Maiduguri, Borno State, Nigeria, an institution previously attacked by six Boko Haram gunmen.
That Dragon Cancer	2016		game	An autobiographical game based on the Green family's experience raising their son Joel, diagnosed with terminal cancer at twelve months old and, though only given a short time to live, continued to survive for four more years.
Things I Cannot Change	1967		film	A NFB film, considered to be an impetus for the NFB's Challenge for Change initiative, that provoked debates on the ethics of documentary. The film harmfully exposed a family living in poverty as they reflected on their experience in Montreal over the course of three weeks.

(continued)

Title	Start Year	End Year	Type of Media	Description
Twine	2009	ongoing	tool	An open-source tool for telling interactive, nonlinear stories.
Two Towns of Jasper	2002		film	The work of a Black film crew and a White film crew in Jasper, Texas, documenting the aftermath of the murder of James Byrd Jr., a Black man chained to moving a vehicle and dragged to death by three White men.
Video nas Aldeias (Video in the Villages)	1986	ongoing	film collection	A pioneer project in the field of Indigenous audiovisual production in Brazil. Since its inception, the goal of the project has been to support Indigenous peoples' struggles in order to strengthen identities and territorial and cultural heritages.
VTR St-Jacques	1969		film	A short film experiment in community engagement and social action. A committee from an impoverished neighborhood in Montreal recorded residents discussing their concerns and then played these recordings for the community at large, sparking collective problem solving.
Walking with Our Sisters	2012	ongoing	textile	A commemorative art installation to honor the lives of missing and murdered Indigenous women of Canada and the United States. The installation is composed of entirely crowd-sourced "vamps," the upper beaded section of moccasins.
Wall, The	2018		platform	A Pulitzer Prize–winning *USA Today Network* investigation involving thirty reporters and photographers. The project explores the US-Mexican border, shedding light on the human, ecological, and economic impacts of a border wall.
wampum.codes	2019	ongoing	audio, performance, tool	Amelia Winger-Bearskin's podcast and design tool to build ethical dependencies into our cultural and technological codes, framed in Indigenous epistemologies.
War of the Worlds	1938		audio	Orson Welles's fake radio documentary that terrified audiences into believing that Martians had invaded New Jersey.
We Feel Fine	2006	2009	data visualization	Jonathan Harris and Sep Kamvar's data-driven project documenting and visualizing emotional data, pulling from blogs and including metadata on weather, author demographics, and author location.

Title	Start Year	End Year	Type of Media	Description
We Live in An Ocean of Air	2018		VR	This VR project takes viewers on a journey, following their breath out from their body through a virtual ecosystem.
Welcome to Chechnya	2020		film	In this observational documentary, director David France goes inside the Chechen Republic's war on the LGBTQ+ community and protects the identities of twenty-six characters in the film, using deepfake technologies to hide their identities.
Wide Awakes, The	2020		performance	An open-source network that radically reimagine the future through creative collaboration.
Wizard of Oz	1939		film	A children's fantasy film in which an all-powerful wizard is revealed to be a mere human hiding behind a curtain and speaking into a sound system.
Women in America Earn Only 2/3 of What Men Do	1985		installation	Appropriating the visual language of advertising, the Guerilla Girls' exhibit critiqued gendered inequity in the art world and the larger economy.
You Are on Indian Land	1969		film	A documentary short by the Indian Film Crew, the NFB's first all-Indigenous film unit, and part of the Challenge for Change initiative. The film documents a protest over actions by the Canadian government to block duty-free border crossings of personal purchases in which Kanien'kéhaka activists blocked the bridge between Ontario and New York.
Zero Days VR	2017		VR	Scatter's Emmy-winning VR visualizes the story of Stuxnet in a new way, placing users inside the invisible world of computer viruses and experiencing the high stakes of cyberwarfare at a human scale.

NOTES

INTRODUCTION

1. Chimamanda Ngozi Adichie, "The Danger of a Single Story," TEDGlobal Conference, Oxford, UK, July 2009, https://www.ted.com/talks/chimamanda_adichie_the_danger_of_a_single_story/transcript?language=en.

2. Ian Sample, "The Great Project: How Covid Changed Science Forever," *The Guardian*, December 15, 2020, https://www.theguardian.com/world/2020/dec/15/the-great-project-how-covid-changed-science-for-ever.

3. Charlie Campbell, "Exclusive: The Chinese Scientist Who Sequenced the First COVID-19 Genome Speaks Out about the Controversies Surrounding His Work," *TIME*, August 24, 2020, https://time.com/5882918/zhang-yongzhen-interview-china-coronavirus-genome.

4. Walter Isaacson, "CRISPR Rivals Put Patents Aside to Help in Fight against Covid-19," *STAT News*, March 3, 2021, https://www.statnews.com/2021/03/03/crispr-rivals-put-patents-aside-fight-against-covid-19/?utm_source=STAT+Newsletters&utm_campaign=caa583f0de-MR_COPY_14&utm_medium=email&utm_term=0_8cab1d7961-caa583f0de-152446341.

5. Walter Isaacson, *The Code Breaker: Jennifer Doudna, Gene Editing, and the Future of the Human Race* (New York: Simon & Schuster, 2021): 473.

6. The term *surveillance capitalism* was popularized by Shoshana Zuboff in her book *The Age of Surveillance Capitalism* (2018) but was coined in 2014 by Canada-based sociologist Vincent Mosco.

7. Duke Redbird, *Dish with One Spoon*, filmed November 2020 at Open Documentary Lab Lecture Series, Cambridge, MA, video, http://opendoclab.mit.edu/presents/duke-redbird-indigenous-digital-delegation.

8. Pierre Levy, *Collective Intelligence: Mankind's Emerging World in Cyberspace* (New York: Basic Books, 1999).

9. Will Fitzgibbon, "What Happened after the Panama Papers?," *International Consortium of Investigative Journalists*, April 3, 2019, https://www.icij.org/investigations/panama-papers/what-happened-after-the-panama-papers/.

CHAPTER 1

1. Katerina Cizek and William Uricchio, with coauthors Juanita Anderson, Maria Aqui Carter, Detroit Narrative Agency, Thomas Allen Harris, Maori Karmael Holmes, Richard Lachman, Louis Massiah, Cara Mertes, Sara Rafsky, Michèle Stephenson, Amelia Winger-Bearskin, and Sarah Wolozin, *Collective Wisdom: Co-Creating Media within Communities, across Disciplines and with Algorithms* (Cambridge, MA: MIT Press), https://wip.mitpress.mit.edu/collectivewisdom. Unless otherwise indicated, quotations are from *Collective Wisdom*.

2. Merrill De Voe, *How to Tailor Your Sales Organization to Your Markets* (New York: Prentice Hall, 1964), 102.

3. T. F. D. Williams and M. Oakescott, *The Cambridge Journal* 7 (1953): 314.

4. "Technology as a Creative Medium: Supporting Artists Who Use Digital Tools to Illuminate Our World," National Endowment of the Arts, forthcoming.

5. Marcia Nickerson, *On Screen Protocols & Pathways: A Media Production Guide to Working with First Nations, Métis and Inuit Communities, Cultures, Concepts, and Stories* (Toronto: imagineNATIVE, 2019), 19.

6. OED Online, December 2018, s.v. "bee," http://www.oed.com/view/Entry/16941.

7. Michael Schudson, *Discovering the News: A Social History of American Newspapers* (New York: Dover Books, 1978).

8. Augusto Boal, "World Theatre Day Message," Speech, Paris 2009, International Theatre Institute, https://www.world-theatre-day.org/pdfs/WTD_Boal_2009.pdf.

9. Henry Jenkins, *Textual Poachers: Television Fans and Participatory Culture* (New York: Routledge, 1992).

10. Third Space is the concept developed by E. W. Soja, *Thirdspace: Journeys to Los Angeles and Other Real-and-Imagined Places* (Oxford, UK: Blackwell Publishing, 1996) and later by Homi Bhabha, *The Location of Culture* (Abingdon, UK: Routledge, 1994).

11. Marina Sitrin, "Definitions of Horizontalism and Autonomy," *NACLA Report on the Americas* 47, no. 3 (2014): 44–45.

12. Joshua Keating, "The George Floyd Protests Show Leaderless Movements Are the Future of Politics," *Slate*, June 9, 2020, https://slate.com/news-and-politics/2020/06/george-floyd-global-leaderless-movements.html.

13. Lam Thuy Vo, "Adversarial Experiments Everywhere," *Creative Futures*, Ford Foundation, https://www.fordfoundation.org/campaigns/creative-futures/?popup=14999.

14. Kamal Sinclair, "The High Stakes of Limited Inclusion," *Making a New Reality*, November 29, 2017, https://makinganewreality.org/the-high-stakes-of-limited-inclusion-8c11541829f6.

15. "Values Statements," *New Public*, https://newpublic.org/purpose/values-statements.

16. Henry Jenkins, *By Any Media Necessary: The New Youth Activism* (New York: Routledge, 2016), 29.

17. Carlos Martínez de la Serna, "Collaboration and the Creation of a New Journalism Commons." Columbia Journalism Review, March 30, 2018, https://www.cjr.org/tow_center_reports /collaboration-and-the-journalism-commons.php.

18. danah boyd, *Strengthening Local News, Media and Democracy*, filmed February 2019 at Knight Media Forum, Miami, FL, video, https://www.knightfoundation.org/kmf-2019.

19. American Journalism Project, http://www.theajp.org.

20. Leyland Cecco, "Google Affiliate Sidewalk Labs Abruptly Abandons Toronto Smart City Project," *The Guardian*, May 7, 2020, https://www.theguardian.com/technology/2020/may/07/google -sidewalk-labs-toronto-smart-city-abandoned.

21. Zephyr Teachout and Pat Garofalo, "Cuomo Is Letting Billionaires Plan New York's Future. It Doesn't Have to Be This Way," *The Guardian*, May 14, 2020, https://www.theguardian.com /commentisfree/2020/may/14/andrew-cuomo-bill-gates-eric-schmidt-coronavirus.

22. Kamal Sinclair, "Design Public Space," *Making a New Reality*, June 1, 2018, https://makinganewreality .org/design-public-space-228ab32965be.

23. C. K. Prahalad and Venkat Ramaswamy, *The Future of Competition: Co-Creating Unique Value with Customers* (Cambridge, MA: Harvard Business School Press, 2004), 8.

24. "Design Justice Network Principles," Design Justice Network, Summer 2018, https:// designjustice.org/read-the-principles.

CHAPTER 2

1. Mandy Rose, "Making Publics: Documentary as Do-it-with-others Citizenship," in *DIY Citizenship: Critical Making and Social Media*, ed. Mat Ratto and Megan Boler (Cambridge: MIT Press, 2014), 2.

2. Rose, 1.

3. Forsyth Hardy, ed., *Grierson on Documentary* (Berkeley: University of California Press, 1971), 148.

4. Dziga Vertov, "Kinopravda & Radiopravda (By Way of Proposal)," in *Kino Eye: The Writings of Dziga Vertov*, ed. Annette Michelson (Berkeley: University of California Press, 1984), 114.

5. Jorge Sanjinés, "Problems of Form and Content in Revolutionary Cinema," in *Film Manifestos and Global Cinema Cultures: A Critical Anthology*, ed. Scott MacKenzie (Berkeley: University of California Press, 2014), 289.

6. Sanjinés, 288.

7. Patricia Aufderheide, *The Daily Planet: A Critic on the Capitalist Culture Beat* (Minneapolis: University of Minnesota Press, 2000).

8. Patricia Zimmermann and Helen De Michiel, *Open Space New Media Documentary: A Toolkit of Theory and Practice* (London: Routledge, 2018).

9. Bill Nemtin and Colin Low, *Fogo Island Film and Community Development Project,* National Film Board of Canada, May 1968, http://onf-nfb.gc.ca/medias/download/documents/pdf/1968-Fogo -Island-Project-Low-Nemtin.pdf.

10. Jeff Webb, "Remembering Colin Low and the Fogo Process," *The Acadenis Blog* (blog), March 29, 2016, https://acadiensis.wordpress.com/2016/03/29/remembering-colin-low-and-the-fogo-process.

11. Zimmerman and De Michiel, *Open Space New Media Documentary*, xviii.

12. Lina Srivastava, "Born into Brothels, Five Years Later," *Context | Culture | Collaboration: Strategy for Social Impact* (blog), December 2009, https://linasrivastava.blogspot.com/2009/12/born-into-brothels-five-years-later.html.

13. Pooja Rangan, *Immediations: The Humanitarian Impulse in Documentary* (Durham, NC: Duke University Press, 2017).

14. Elizabeth Miller, *Going Public: The Art of Participatory Practice* (Vancouver: UBC Press, 2017), 44.

15. Paul Vitello, "George C. Stoney, Documentary Filmmaker, Dies at 96," *New York Times*, July 15, 2012, https://www.nytimes.com/2012/07/15/arts/television/george-c-stoney-documentarian-dies-at-96.html.

16. Anthony Riddle and George Stoney, "Local Public Access TV under Attack from Trio of Congressional Bills," interview by Amy Goodman, *Democracy Now!* September 30, 2005, https://www.democracynow.org/2005/9/30/local_public_access_tv_under_attack.

17. Okwui Enwezor, "The Artist as Producer in Times of Crisis," Dark Matter, n.d., http://www.darkmatterarchives.net/wp-content/uploads/2011/02/Enwezor.AuthorProd.pdf.

18. Truth and Reconciliation Commission of Canada, *Honouring the Truth, Reconciling for the Future: Summary of the Final Report of the Truth and Reconciliation Commission of Canada* (Winnipeg, Canada: National Center for Truth and Reconciliation, 2015), https://ehprnh2mwo3.exactdn.com/wp-content/uploads/2021/01/Executive_Summary_English_Web.pdf.

19. Adam DePollo, "Riding with Aunt Dot: An Interview with Detroit Artist Bree Gant," *Michigan Quarterly Review Online* (blog), March 7, 2018, https://sites.lsa.umich.edu/mqr/2018/03/riding-with-aunt-ddot-an-interview-with-detroit-artist-bree-gant.

20. Leo Mirani, "Millions of Facebook Users Have No Idea They're Using the Internet," *Quartz*, February 9, 2015, https://qz.com/333313/milliions-of-facebook-users-have-no-idea-theyre-using-the-internet.

21. Aiman Khan and Ishita Chakrabarty, "Why the 2020 Violence in Delhi Was a Pogrom," Al Jazeera, February 24, 2021, https://www.aljazeera.com/opinions/2021/2/24/why-the-2020-violence-in-delhi-was-a-pogrom.

22. Paul Mozur, "A Genocide Incited on Facebook, with Posts from Myanmar's Military," *New York Times*, October 15, 2018, https://www.nytimes.com/2018/10/15/technology/myanmar-facebook-genocide.html.

23. Karen Hao, "How Facebook Got Addicted to Spreading Misinformation," MIT *Technology Review*, March 11, 2021, https://www.technologyreview.com/2021/03/11/1020600/facebook-responsible-ai-misinformation/?truid=3ffc0dbdcf40b18a84d42e3a770bddeb&utm_source=the_download&utm_medium=email&utm_campaign=the_download.unpaid.engagement&utm_term=Active%20Qualified&utm_content=03-11-2021&mc_cid=a5caea9a4f&mc_eid=2686b6ffcf.

24. Joseph Johnson, "Global Digital Population as of January 2021," *Statista*, March 5, 2021, https://www.statista.com/statistics/617136/digital-population-worldwide/#:~:text=As%20of%20January%202021%20there,the%20internet%20via%20mobile%20devices.

25. Robert J. Deibert, *Reset: Reclaiming the Internet for Civil Society* (Toronto: House of Anansi Press, 2020), 275.

26. "Reach.love," *Docubase*, n.d., https://docubase.mit.edu/tools/reach.

27. "VoIP Drupal," Drupal, November 15, 2018, https://www.drupal.org/project/voipdrupal.

28. "Proyecto Quipu: El drama de las esterilizaciones forzadas en la dictadura de Fujimori," *The Clinic*, February 28, 2016, https://www.theclinic.cl/2016/02/28/proyecto-quipu-el-drama-de-las -esterilizaciones-forzadas-en-la-dictadura-de-fujimori.

29. "The Quipu Project: A Framework for Participatory Interactive Documentary," i-Docs, n.d., http://i-docs.org/the-quipu-project-a-framework-for-participatory-interactive-documentary.

30. Mandy Rose, "Not Media About, but Media With: Co-Creation for Activism," in *I-Docs: The Evolving Practices of Interactive Documentary*, ed. Judith Aston, Sandra Gaudenzi, and Mandy Rose (New York: Columbia University Press, 2017), 58.

31. Deborah Willis and Natasha L. Logan, *Question Bridge: Black Males in America* (New York: Aperture Foundation and Campaign for Black Male Achievement, 2015).

32. "Case Study: QuestionBridge Black Males," Media Impact Funders, January 2012, https:// mediaimpactfunders.org/wp-content/uploads/2016/06/QuestionBridge.pdf.

33. Andrew DeVigal, "The Continuum of Engagement: Pushing the Boundaries of News Engagement Day," *Let's Gather*, October 3, 2017, https://medium.com/lets-gather/the-continuum-of -engagement-89778f9d6c3a.

34. "Collaborative Journalism Database," Center for Cooperative Media, n.d., https://collaborative journalism.org/database-search-sort-learn-collaborative-projects-around-world.

35. Sarah Stonbeley, *Comparing Models of Collaborative Journalism* (Montclair, NJ: Center for Cooperative Media, September 2018).

36. Robert Gebeloff, Danielle Ivory, Matt Richtel, Mitch Smith, Karen Yourish, Jackie Fortiér, Elly Yu, and Molly Parker, "The Striking Racial Divide in How Covid-19 Has Hit Nursing Homes," *New York Times*, September 10, 2020, https://www.nytimes.com/article/coronavirus-nursing-homes -racial-disparity.html.

37. Project Facet, http://www.projectfacet.org.

38. William Uricchio and Sarah Wolozin, *Mapping the Intersections of Two Cultures: Interactive Documentary and Digital Journalism* (Cambridge, MA: MIT Open Documentary Lab, 2016), http:// opendoclab.mit.edu/interactivejournalism/Mapping_the_Intersection_of_Two_Cultures_Interac tive_Documentary_and_Digital_Journalism.pdf.

39. Tal Lorberbaum, Kevin J. Sampson, Raymond L. Woosley, Robert S. Kass, and Nicholas P. Tatonetti, "An Integrative Data Science Pipeline to Identify Novel Drug Interactions That Prolong the QT Interval," *Drug Safety* 39 (May 2016): 433–444.

40. Sam Roe, "Could Collaborating with Scientists Be the Next Step for Investigative Reporting?," *Columbia Journalism Review*, February 12, 2016, https://www.cjr.org/first_person/could_collabo rating_with_scientists_be_the_next_step_for_investigative_reporting.php.

41. Ben DeJarnette, "Journalists and Academics Collaborating? It's Paying Off for Investigative Reports in Canada," *MediaShift* (blog), July 1, 2016, http://mediashift.org/2016/07/journalists -academics-collaborating-its-paying-off-investigative-reports-canada.

42. "Journalists, Academics Discuss Collaboration in 'Mind to Mind' Symposium," *Reveal* (blog), October 23, 2017, https://revealnews.org/press/journalists-academics-discuss-collaboration-in-mind -to-mind-symposium.

43. "About Education Lab," *Seattle Times*, n.d., https://www.seattletimes.com/education-lab-about.

44. Jihii Jolly, "Platform Aimed at Audience Interaction Generates Story Ideas, Goodwill," *Columbia Journalism Review*, August 22, 2016, https://www.cjr.org/the_profile/hearken_hey_area_homeless _san_francisco_audience.php?utm_content=bufferc4c9f&utm_medium=social&utm_source=twitter .com&utm_campaign=buffer.

45. SecureDrop, https://securedrop.org.

46. Joy Mayer, "It's Time to Gather, and We Hope You'll Join Us," *Let's Gather*, October 1, 2017, https://medium.com/lets-gather/its-time-to-gather-and-we-hope-you-ll-join-us-85700c4a9fe3.

47. Ruha Benjamin, *Race after Technology: Abolitionist Tools for the New Jim Code* (Hoboken, NJ: Wiley, 2019), 17.

48. Duke Redbird, *Dish with One Spoon*, video filmed November 2020 at the Open Documentary Lab Lecture Series, Cambridge, MA, http://opendoclab.mit.edu/presents/duke-redbird-indigenous -digital-delegation.

49. Tim Berners-Lee, "The Web Is under Threat. Join Us and Fight for It," *World Wide Web Foundation* (blog), March 12, 2018, https://webfoundation.org/2018/03/web-birthday-29.

50. Ruha Benjamin, *Race after Technology*, 192.

51. Redbird, *Dish with One Spoon*.

CHAPTER 3

1. Evgeny Morozov, *To Save Everything, Click Here: The Folly of Technological Solutionism* (New York: Public Affairs, 2013).

2. Karen Hao, "How Facebook Got Addicted to Spreading Misinformation," *Technology Review*, March 11, 2021, https://www.technologyreview.com/2021/03/11/1020600/facebook-responsible-ai -misinformation.

3. Kamal Sinclair, "Share Space," *Making a New Reality*, March 18, 2018, https://makinganewreality .org/share-space-fc4ff7ae6d5e.

4. Stuart Jeffries, "When Two Tribes Meet: Collaborations Between Artists and Scientists," *The Guardian*, August 21, 2011, https://www.theguardian.com/artanddesign/2011/aug/21/collaborations -between-artists-and-scientists.

5. Gina Czarnecki, "On Our Collaboration," Gina Czarnec (blog), January 5, 2018, https://www .ginaczarnecki.com/single-post/2018/01/05/On-our-collaboration.

6. "About," iSeeChange, n.d., https://www.iseechange.org/about/mission.

7. This was a ministry in the British government between 1968 and 1988.

8. Nora Kahn, "Fred Turner: Silicon Valley Thinks Politics Doesn't Exist," 032c, July 30, 2018, https://032c.com/fred-turner-silicon-valley-thinks-politics-doesnt-exist.

9. Fred Turner, *From Counterculture to Cyberculture: Stewart Brand, the Whole Earth Network and the Rise of Digital Utopianism* (Chicago: University of Chicago Press, 2006).

10. Gisela Williams, "Are Artists the Interpreters of New Scientific Innovation?," *New York Times Magazine*, September 12 2017, https://www.nytimes.com/2017/09/12/t-magazine/art/artist-residency -science.html.

11. Vanessa Chang, "Recoding the Master's Tools: Artists Remake Systems of Oppression and Extraction in Technology," National Endowment for the Arts, https://www.arts.gov/impact /media-arts/arts-technology-scan/essays/recoding-masters-tools-artists-remake-systems-oppression -and-extraction-technology.

12. "REACT Report 2012–2016," n.d., Research & Enterprise in Arts and Creative Technology, http:// www.react-hub.org.uk/sites/default/files/publications/REACT%20Report%20low%20res_2.pdf.

13. "Watershed Sandbox: A How To Guide," Watershed, n.d., https://www.watershed.co.uk/sites /default/files/publications/2018-10-08/watershedsandbox_ahowtoguide.pdf.

14. "Values," Eyebeam, n.d., https://www.eyebeam.org.

15. Vanessa Chang, *The Grid Art + Tech Report 2020*, n.d., commissioned by EUNIC Silicon Valley, https://09227776-4c3f-45f5-9037-85b84a0e59af.filesusr.com/ugd/f31f46_8151ab52dbe040a6af559d c574be57d3.pdf?fbclid=IwAR2BTgInAbg4LKPLNAhVGvc7SG86Pt-GxuuJL6NZ0n5qUthMBwM5ie _I0L8.

16. Third Space is a concept developed in E. W. Soja, *Thirdspace: Journeys to Los Angeles and Other Real-and-Imagined Places* Oxford, UK: Blackwell Publishing, 1996), and later in Homi Bhabha, *The Location of Culture* (Abingdon, UK: Routledge, 2004).

17. "New Frontiers in Research Fund," Government of Canada, November 27, 2020, https://www .sshrc-crsh.gc.ca/funding-financement/nfrf-fnfr/index-eng.aspx.

18. Joichi Ito, "Antidisciplinary," Joi Ito (blog), October 2, 2014, https://joi.ito.com/weblog/2014 /10/02/antidisciplinar.html.

19. Hyphen-Labs, http://www.hyphen-labs.com.

20. "NSAF," Hyphen-Labs, March 23, 2021, http://www.hyphen-labs.com/nsaf.html.

21. Sameer Rao, "How 'Neurospeculative AfroFeminism' Uses Virtual Reality to Explore Other-worldly Transformation at a Black Hair Salon," *Colorlines*, March 17, 2017, https://www.colorlines .com/articles/how-neurospeculative-afrofeminism-uses-virtual-reality-explore-otherworldly -transformation.

22. Sue Ding, "Interview with Hyphen-Labs," Docubase, n.d., https://docubase.mit.edu/lab /interviews/by-htm.

23. Catherine Porter, "Reviving a Lost Language of Canada through Film," *New York Times*, June 11, 2017, https://www.nytimes.com/2017/06/11/world/americas/reviving-a-lost-language-of-canada -through-film.html.

24. *Council of the Haida Nation v. Province of British Columbia and Canada*, No. L020662 (Vancouver, 2002).

25. "History of the Haida Nation," Haida Nation, n.d., https://www.haidanation.ca/?page_id=26.

26. Leonie Sandercok, Dana Moraes, and Jonathan Frantz, "Film as a Catalyst for Indigenous Community Development," *Planning Theory and Practice* 18, no. 4: 639–666.

27. Maryse Zeidler, "Inuit Film Producers Partner with Haida for First-of-Its-Kind Feature," *CBC Arts*, January 15, 2017, https://www.cbc.ca/news/canada/british-columbia/the-fast-runner-haida -film-1.3935746.

CHAPTER 4

1. Margaret A. Boden, *AI: Its Nature and Future* (Oxford: Oxford University Press, 2016); Alan Turing, "Computing Machinery and Intelligence," *Mind* 59, no. 236 (1950): 433–460.

2. Kate Crawford, *Atlas of AI: Power, Politics and the Planetary Costs of Artificial Intelligence* (New Haven, CT: Yale University Press, 2021), 10–11.

3. Kate Darling, *The New Breed: What Our History with Animals Reveals about Our Future with Robots* (New York: Henry Holt, 2021), xiv.

4. Colin Allen and Michael Trestman, "Animal Consciousness," *The Stanford Encyclopedia of Philosophy Archive*, Winter 2017, https://plato.stanford.edu/archives/win2017/entries/consciousness-animal.

5. C. Dianne Martin, "ENIAC: The Press Conference That Shook the World," *IEEE Technology and Society Magazine*, December 1995.

6. "New Navy Device Learns by Doing; Psychologist Shows Embryo of Computer Designed to Read and Grow Wiser," *New York Times*, July 8, 1958, 25.

7. Lovelace's comment referred to Charles Babbage's Analytical Engine, a device conceptualized in the same decade as photography and telegraphy, making the 1830s foundational to our current media order.

8. Steve Lohr, "M.I.T. Plans College for Artificial Intelligence, Backed by $1 Billion," *New York Times*, October 15, 2018, https://www.nytimes.com/2018/10/15/technology/mit-college-artificial-intelligence.html. For an article criticizing the "hysteria about the future of artificial intelligence," see Rodney Brooks, "The Seven Deadly Sins of AI Predictions." *MIT Technology Review*, October 6, 2017.

9. Jason Edward Lewis, Noelani Arista, Archer Pechawis, and Suzanne Kite, "Making Kin with the Machines," *Journal of Design and Science*, July 16, 2018, https://jods.mitpress.mit.edu/pub/lewis-arista-pechawis-kite/release/1.

10. Joichi Ito, "Resisting Reduction: A Manifesto," *Journal of Design and Science*, October 13, 2017, https://jods.mitpress.mit.edu/pub/resisting-reduction/release/17.

11. "Aganetha Dyck Reveals How She Works with Bees to Create Strange and Wonderful Art," *CBC Arts*, March 29, 2018, https://www.cbc.ca/arts/aganetha-dyck-reveals-how-she-works-with-bees-to-create-strange-and-wonderful-art-1.4597098.

12. "Aganetha Dyck Reveals How She Works with Bees."

13. A generative adversarial network is a system of two competing neural networks in which one generates content and the other evaluates the content. As more data is input, the networks improve in their pattern recognition.

14. Kristin Houser, "AI-Generated Art Will Go on Sale alongside Human-Made Works This Fall," *Futurism*, August 22, 2018, https://futurism.com/the-byte/ai-generated-portrait-christies. See also "Is Artificial Intelligence Set to Become Art's Next Medium?," *Christie's*, December 12, 2018, https://www.christies.com/features/A-collaboration-between-two-artists-one-human-one-a-machine-9332-1.aspx.

15. Stroud Cornock and Ernest Edmonds, "The Creative Process Where the Artist Is Amplified or Superseded by the Computer," *Leonardo* 6, no. 1 (Winter 1973): 11–16.

16. Sequences of possible events in which the probability of each event depends only on the state attained in the previous event, Markov chains are often used as statistical models of real-world processes.

17. shirin anlen, "When Machines Look for Order in Chaos," *Immerse*, September 4, 2018, https://immerse.news/when-machines-look-for-order-in-chaos-198fb222b60a.

18. For poetry, see Bei Lui, Jianlong Fu, Makoto P. Kato, and Masatoshi Yoshikawa, "Beyond Narrative Description: Generating Poetry from Images by Multi-Adversarial Training," in 2018 ACM Multimedia Conference (MM '18), October 22–26, 2018, Seoul, Republic of Korea, https://doi.org/10.1145/3240508.3240587. For literature, see Yurei Raita, *The Day a Computer Writes a Novel* (n.p., 2016), an AI-generated novel that nearly won the Nikkei Hoshi Shinichi Literary Award in 2016. For movies, see *Sunspring* (Oscar Sharp, 2016), an AI-generated sci-fi short film. For theater, see Simon Colton's musical theater piece *Beyond the Fence* (directed by Luke Sheppard and performed in London's Arts Theater, 2016).

19. Alex Hern, "New AI Fake Text Generator May Be Too Dangerous to Release, Say Creators," *The Guardian*, February 14, 2019, https://www.theguardian.com/technology/2019/feb/14/elon-musk-backed-ai-writes-convincing-news-fiction?fbclid=IwAR3IEShRs0kUW8-h29qCunefcMz62CQOhzFYKzx7oSVnHh4NaZWkYvhohrY.

20. Corinna Underwood, "Automated Journalism—AI Applications at New York Times, Reuters, and Other Media Giants," *Emerj*, November 17, 2019, https://emerj.com/ai-sector-overviews/automated-journalism-applications.

21. Peter Haff, "Humans and Technology in the Anthropocene: Six Rules," *Anthropocene Review* 1, no. 2 (August 1, 2014): 126–136.

22. Tim Adams, "Trevor Paglen: Art in the Age of Mass Surveillance," *The Guardian*, November 25, 2017, https://www.theguardian.com/artanddesign/2017/nov/25/trevor-paglen-art-in-age-of-mass-surveillance-drones-spy-satellites.

23. SITU Research, "Euromaidan Event Reconstruction," Carnegie Mellon University, Center for Human Rights, 2018, https://situ.nyc/research/projects/euromaidan-event-reconstruction.

24. Jackson Ayres, "Orson Welles's 'Complicitous Critique': Postmodern Paradox in 'F for Fake,'" *Literature/Film Quarterly* 40, no. 1 (2021): 6–19.

25. Helen Rosner, "A Haunting New Documentary about Anthony Bourdain," *New Yorker*, July 15, 2021, https://www.newyorker.com/culture/annals-of-gastronomy/the-haunting-afterlife-of-anthony-bourdain.

26. "Prepare, Don't Panic: Synthetic Media and Deepfakes," Witness Media Labs (blog), n.d., https://lab.witness.org/projects/synthetic-media-and-deep-fakes.

27. Sherry Turkle, "There Will Never Be an Age of Artificial Intimacy," *New York Times*, August 11, 2018, https://www.nytimes.com/2018/08/11/opinion/there-will-never-be-an-age-of-artificial-intimacy.html.

28. Simon Parkin, "The YouTube Stars Heading for burnout: 'The Most Fun Job Imaginable Became Deeply Bleak,'" *The Guardian*, September 8, 2018, https://www.theguardian.com/technology/2018/sep/08/youtube-stars-burnout-fun-bleak-stressed?CMP=Share_iOSApp_Other.

29. Kevin McNally, "It's Time to Stop Using AI as a Marketing Gimmick," *Fast Company*, July 18, 2017, https://www.fastcompany.com/90133598/its-time-to-stop-using-ai-as-a-marketing-gimmick.

30. Astra Taylor, "The Automation Charade," *Logic*, August 1, 2018, https://logicmag.io/failure/the-automation-charade.

31. George Joseph and Kenneth Lipp, "IBM Used NYPD Surveillance Footage to Develop Technology That Lets Police Search by Skin Color," *The Intercept*, September 6, 2018, https://theintercept.com/2018/09/06/nypd-surveillance-camera-skin-tone-search.

32. Joy Buolamwini "Hearing on Facial Recognition Technology (part 1): Its Impact on Our Civil Rights and Liberties," written testimony for the United States House Committee on Oversight and Government Reform, May 22, 2019, https://www.congress.gov/116/meeting/house/109521/witnesses/HHRG-116-GO00-Wstate-BuolamwiniJ-20190522.pdf.

33. Sasha Costanza-Chock, "Design Justice, A.I., and Escape from the Matrix of Domination," *Journal of Design and Science*, July 16, 2018, https://jods.mitpress.mit.edu/pub/costanza-chock/release/4.

34. The organizations Costanza-Chock named in our interview are Data & Society, the A.I. Now Institute, and the Digital Equity Lab in New York City; the new Data Justice Lab in Cardiff; and the Public Data Lab. Coding Rights, led by hacker, lawyer, and feminist Joana Varon, works across Latin America to make complex issues around data and human rights much more accessible for broader publics, engage in policy debates, and help produce consent culture for the digital environment. They do this through projects like Chupadatos ('the data sucker'). Other groups include Fair Algorithms; the Data Active group; the Center for Civic Media at MIT; the Digital Justice Lab, recently launched by Nasma Ahmed in Toronto; Building Consentful Tech, by the design studio And Also Too in Toronto; the Our Data Bodies project, by Seeta Ganghadaran and Virginia Eubanks; and the FemTechNet network" from their "Design Justice."

35. William Uricchio, "Recursive Media," *Questions de communication* (forthcoming in French and English, 2022).

36. Stephanie Dinkins, "Conversations with Bina48," 2014–ongoing, https://www.stephaniedinkins.com/conversations-with-bina48.html.

37. Stephanie Dinkins, "Afro-now-ism," *Noēma*, June 16, 2020, https://www.noemamag.com/afro-now-ism.

38. Dinkins, "Afro-now-isms."

CHAPTER 5

1. Tabitha Jackson, *The Art of Co-Creation: A Storytelling Model for Impact and Engagement*, YouTube, April 13, 2018, https://www.youtube.com/watch?v=ALCwYtD2x3E&ab_channel=Skoll.org.

2. Jo Freeman, "The Tyranny of Structurelessness," *Second Wave* 2, no. 1 (1972): 20–33.

3. Willis himself has come in for criticism for his unsanctioned remix of a famous photo by South African photographer Graeme Williams, which could be interpreted as a case of co-creation gone awry. Hank Eileen Kinsella, "Willis Thomas Pulls Work from a South African Art Fair after a Photographer Levels Plagiarism Charges," Artnet, September 13, 2018, https://news.artnet.com/art-world/artist-hank-willis-thomas-pulls-work-from-art-fair-after-photographer-levels-plagiarism-charges-1347710.

4. Bassey Etim, "The Times Sharply Increases Articles Open for Comments, Using Google's Technology," *New York Times*, June 13, 2017, https://www.nytimes.com/2017/06/13/insider/have-a-comment-leave-a-comment.html.

5. James Vincent, "Twitter Taught Microsoft's AI Chatbox to be a Racist Asshole in Less than a Day," *The Verge*, March 24, 2016, https://www.theverge.com/2016/3/24/11297050/tay-microsoft-chatbot-racist.

6. "Platform Co-op Resource Library," Platform Cooperativism Consortium, n.d., https://resources.platform.coop.

7. Ruha Benjamin, *Race after Technology: Abolitionist Tools for the New Jim Code* (Hoboken, NJ: Wiley, 2019), 17.

8. Babitha George, "Making Space for the Informal," *Reflecting on Toolkits*, March 9, 2016, https://medium.com/reflecting-on-toolkits/making-space-for-the-informal-99f0968d1bc0.

9. Nora Khan, "Fred Turner: Silicon Valley Thinks Politics Doesn't Exist," *032c*, July 30, 2018, https://032c.com/fred-turner-silicon-valley-thinks-politics-doesnt-exist.

10. Gloria Steinem, *My Life on the Road* (New York: Penguin Random House, 2016), 16.

11. Andrew Lowenthal, Tanya Notley, Lina Srivastava, and Egbert Wits, "Video for Change Impact Toolkit," *Video for Change*, 2019 https://toolkit.video4change.org.

12. Dziga Vertov, "We: Variant of a Manifesto," *Kino-Fot* (August 1922): 5; The Mentor, "Hacker's Manifesto," *Phrack Magazine* 7 (September 25, 1986), http://phrack.org/issues/7/3.html; Martine Syms, "The Mundane Afrofuturist Manifesto," *Rhizome*, December 17, 2013, https://rhizome.org/editorial/2013/dec/17/mundane-afrofuturist-manifesto.

13. Laura Cutter, "Walter Reed, Yellow Fever, and Informed Consent," *Military Medicine* 181, no. 1 (2016): 90–91.

14. Terri Janke, *Pathways & Protocols: A Filmmaker's Guide to Working with Indigenous People, Culture and Concepts* (n.p.: Australian Government, Screen Australia, 2009), 4.

15. Marcia Nickerson, *On-Screen Protocols & Pathways: A Media Production Guide to Working with First Nations, Métis and Inuit Communities, Cultures, Concepts and Stories* (Toronto: imagiNative, 2019).

16. "Community Benefits 101," *The Parkdale People's Economy Project* (blog), July 28, 2017, https://parkdalecommunityeconomies.wordpress.com/2017/07/28/community-benefits-101.

17. George Stoney, "George Stoney Visits American University," ed. Michael T. Miller and Maura Ugarte, Center for Social Media & Impact, n.d., https://cmsimpact.org/resource/george-stoney-visits-american-univeristy.

18. Colin Low, "Grierson and Challenge for Change (1984)," in *Challenge for Change: Activist Documentary at the National Film Board of Canada*, ed. Thomas Waugh, Michael Brendan Baker, and Ezra Winton (Montreal: McGill-Queen's University Press, 2010), 16.

19. Alexandra Juhaxz and Alisa Lebow, eds., "Beyond Story: Ways to Engage," in *World Records* 5 (2021), https://vols.worldrecordsjournal.org/05/01?index=1.

20. Patricia R. Zimmermann and Helen De Michiel, *Open Space New Media Documentary: A Toolkit for Theory and Practice* (New York: Routledge, 2018), xx.

21. June Cross, "We, the People," Immerse, July 26, 2019, https://immerse.news/we-the-people-f3947f23c792.

22. Amelia Winger-Bearskin, "Before Everyone Was Talking about Decentralization, Decentralization Was Talking to Everyone," *Immerse*, July 2, 2018, https://immerse.news/decentralized-storytelling-d8450490b3ee.

23. "CPJ's Database on Attacks on the Press," The Committee to Protect Journalists, n.d., https://cpj.org/data/killed/?status=Killed&motiveConfirmed%5B%5D=Confirmed&type%5B%5D=Journalist&start_year=1992&end_year=2021&group_by=year.

24. Thenmozhi Soundararajan, *Socially Engaged Art in Volatile Times*, video filmed February 16, 2021, at the Open Documentary Lab Lecture Series, Cambridge, MA, http://opendoclab.mit.edu/presents/thenmozhi-soundararajan-equality-labs.

25. Marcia Nickerson, *On Screen Protocols & Pathways: A Media Production Guide to Working with First Nations, Métis and Inuit Communities, Cultures, Concepts, and Stories* (Toronto: imagiNative, 2019).

26. adrienne maree brown, *Emergent Strategy: Shaping Change, Changing Worlds* (Chicago: AK Press, 2017), 135.

27. "Creative Futures," Ford Foundation, n.d., https://www.fordfoundation.org/campaigns/creative-futures/?popup=14896.

28. Sam Gregory, "It Gets Better: Collective and Individual Voice in Video Advocacy," *WITNESS* (blog), November 2011, https://blog.witness.org/2010/11/it-gets-better-collective-and-individual-voice-in-video-advocacy.

29. Zeynep Tufekci, *Twitter and Tear Gas* (New Haven, CT: Yale University Press, 2017), xxvii.

30. Benjamin, *Race after Technology*, 42.

31. *Tech as Art: Supporting Artists Who Use Technology as a Creative Medium*, National Endowment for the Arts, 2021, https://www.arts.gov/about/publications/tech-art-supporting-artists-who-use-technology-creative-medium.

32. "Tahir Hemphill Studio," TH Media, n.d., https://www.tahirhemphill.com.

33. Ronald J. Deibert, *Reset: Reclaiming the Internet for Civil Society* (Toronto: House of Anansi Press, 2020), 269.

34. Deibert, 317.

35. Angie Kim, "The Co-Op," Ford Foundation, n.d., https://www.fordfoundation.org/campaigns/creative-futures/?popup=angie-kim.

36. Ruth Catlow, Marc Garrett, Nathan Jones, and Sam Skinner, eds., *Artists Re:Thinking the Blockchain* (Liverpool: Liverpool University Press, 2018), 31.

37. Steven Johnson, *Emergence: The Connected Lives of Ants, Brains, Cities, and Software* (New York: Scribner, 2002).

38. brown, *Emergent Strategy*, 3.

BIBLIOGRAPHY

"About." iSeeChange, n.d., https://www.iseechange.org/about/mission.

"About Education Lab." *Seattle Times*, n.d., https://www.seattletimes.com/education-lab-about.

Adams, Tim. "Trevor Paglen: Art in the Age of Mass Surveillance." *The Guardian,* November 25, 2017, https://www.theguardian.com/artanddesign/2017/nov/25/trevor-paglen-art-in-age-of-mass -surveillance-drones-spy-satellites.

Adichie, Chimamanda Ngozi. "The Danger of a Single Story." TEDGlobal Conference, Oxford, UK, July 2009, https://www.ted.com/talks/chimamanda_adichie_the_danger_of_a_single_story /transcript?language=en.

"Aganetha Dyck Reveals How She Works with Bees to Create Strange and Wonderful Art," CBC Arts, March 29, 2018, https://www.cbc.ca/arts/aganetha-dyck-reveals-how-she-works-with-bees -to-create-strange-and-wonderful-art-1.4597098.

Allen, Colin, and Michael Trestman. "Animal Consciousness." *Stanford Encyclopedia of Philosophy Archive*, October 24, 2016, https://plato.stanford.edu/archives/win2017/entries/consciousness -animal.

American Journalism Project. http://www.theajp.org.

anlen, shirin. "When Machines Look for Order in Chaos." *Immerse,* September 4, 2018, https:// immerse.news/when-machines-look-for-order-in-chaos-198fb222b60a.

Aoun, Joseph E. *Robot Proof: Higher Education in the Age of Artificial Intelligence*. Cambridge, MA: MIT Press, 2017.

Arnstein, Sherry R. "A Ladder of Citizen Participation." *Journal of the American Institute of Planners* 35, no. 4 (1969): 216–224.

Arvind, Narayanan. Slides from a talk at MIT, Princeton University/Center for Information technology Policy, November 18, 2019, https://www.cs.princeton.edu/~arvindn/talks/MIT-STS-AI-snakeoil.

Aston, Judith, and Sandra Gaudenzi. "Interactive Documentary: Setting the Field." *Studies in Documentary Film* 6, no. 2 (2012): 125–139.

Aston, Judith, Sandra Gaudenzi, and Mandy Rose, eds. *I-docs: The Evolving Practices of Interactive Documentary*. New York: Columbia University Press, 2017.

Aufderheide, Patricia. *The Daily Planet: A Critic on the Capitalist Culture Beat*. Minneapolis: University of Minnesota Press, 2000.

Aufderheide, Patricia. "The Video in the Villages Project: Videomaking with and by Brazilian Indians," *Visual Anthropology Review* 11, no. 2 (1995): 83–93.

Auguiste, Reece. "Another Way of Being: Hidden Histories of Collaborative Documentary Practices." *Afterimage: The Journal of Media Arts and Cultural Criticism*, May 15, 2017.

Auguiste, Reece, Helen De Michiel, Aggie Ebrahimi Bazaz, and Patricia R. Zimmermann . "Speculations and Inquiries on New Participatory Documentary Environments." *Afterimage: The Journal of Media Arts and Cultural Criticism*, 2017.

Ayres, Jackson. "Orson Welles's 'Complicitous Critique': Postmodern Paradox in 'F for Fake.'" *Literature-Film Quarterly* 40, no.1 (2021): 6–19.

Banks, John, and Mark Deuze. "Co-Creative Labour," *International Journal of Cultural Studies*. 12, no. 5 (2009): 419–431.

Barad, Karen. *Meeting the Universe Halfway: Quantum Physics and the Entanglement of Matter and Meaning*. Durham, NC: Duke University Press, 2007.

Barbash, Ilisa, and Lucien Taylor. *Cross-cultural Filmmaking: A Handbook for Making Documentary and Ethnographic Films and Videos*. Oakland: University of California Press, 1997.

Barclay, Barry. *Our Own Image: A Story of a Māori Filmmaker*. Minneapolis: University of Minnesota Press, 2015.

Bassel, Leah. *The Politics of Listening*. New York: Palgrave Macmillan, 2017.

Batist, Danielle. "Why Co-Creation Is the Future for All of Us." *Forbes,* February 4, 2014, https://www.forbes.com/sites/ashoka/2014/02/04/why-co-creation-is-the-future-for-all-of-us/#5f9d5d82a653.

Battaglia, Giulia. "Crafting 'Participatory' and 'Collaborative' Film-Projects in India." *Anthrovision* 2, no. 2 (2014): 1–22.

Becker, Howard. "Art as a Collective Action." *American Sociological Review* 39, no. 6 (1974): 767–776, http://citeseerx.ist.psu.edu/viewdoc/download?doi=10.1.1.472.6502&rep=rep1&type=pdf.

Beech, Dave. "Include Me Out!" *Art Monthly*, 315 (2008), http://www.artmonthly.co.uk/magazine/site/article/include-me-out-by-dave-beech-april-2008.

Benjamin, Ruha. *Race after Technology: Abolitionist Tools for the New Jim Code*. Cambridge, UK: Polity, 2019.

Benjamin, Walter. "On the Concept of History." *Walter Benjamin Archive/Writer's Archive/Marxists Internet Archive,* 1940, https://www.marxists.org/reference/archive/benjamin/1940/history.htm.

Benjamin, Walter. "The Work of Art in the Age of Mechanical Reproduction." Trans. Harry Zohn. *Walter Benjamin Archive/Writer's Archive/Marxists Internet Archive,* 1936, https://www.marxists.org /reference/subject/philosophy/works/ge/benjamin.htm.

Bennett, Jane. *Vibrant Matter: A Political Ecology of Things.* Durham, NC: Duke University Press, 2010.

Bernard, Sheila Curran. "Eyes on the Rights: The Rising Cost of Putting History on Screen." *Documentary Magazine*, International Documentary Association, June 2005, https://www.documentary .org/feature/eyes-rights-rising-cost-putting-history-screen.

Bernard, Sheila Curran. "Watching *Eyes on the Prize.*" In *DoubleTake/Points of Entry* (Fall/Winter 2006). https://www.sheilacurranbernard.com/uploads/1/0/2/7/10273986/doubletake-watching_ eyes.pdf.

Berners-Lee, Tim. "The Web Is Under Threat. Join Us and Fight for It," World Wide Web Foundation (blog), March 12, 2018, https://webfoundation.org/2018/03/web-birthday-29.

Berry, Marsha, and Max Schleser, eds. *Mobile Media Making in an Age of Smartphones.* New York: Palgrave Pivot, 2014.

Bhabha, Homi. *The Location of Culture.* Abingdon, UK: Routledge, 1994.

Bianchini, Samuel, and Erik Verhagen, eds. *Practicable: From Participation to Interaction in Contemporary Art.* Cambridge, MA: MIT Press, 2016.

Bishop, Claire. *Participation.* Cambridge, MA: MIT Press, 2006.

Boal, Augusto. *Theatre of the Oppressed.* London: Pluto, 1979.

Boal, Augusto. "World Theatre Day Message, Speech, International Theatre Institute, Paris 2009, https://www.world-theatre-day.org/pdfs/WTD_Boal_2009.pdf.

Boden, Margaret A. *AI: Its Nature and Future.* Oxford: Oxford University Press, 2016.

Boggs, Grace Lee. *Living for Change: An Autobiography.* Minneapolis: University of Minnesota Press, 2016.

Bohm, David. *On Dialogue.* New York: Routledge, 1996.

Boler, Megan, and Matt Ratto, eds. *DIY Citizenship: Critical Making and Social Media.* Cambridge, MA: MIT Press, 2014.

Bourriaud, Nicolas. "Relational Aesthetics." Seth Kim-Cohen, 1998. http://www.kim-cohen.com /seth_texts/artmusictheorytexts/Bourriaud%20Relational%20Aesthetics.pdf.

boyd, danah. *Strengthening Local News, Media and Democracy.* Video filmed February 2019 at Knight Media Forum, Miami, FL, https://www.knightfoundation.org/kmf-2019.

Brabham, Daren C. *Crowdsourcing.* Cambridge, MA: MIT Press, 2013.

Brady, Miranda, and John M. H. Kelly. *We Interrupt This Program: Indigenous Tactics in Canadian Culture.* Vancouver: UBC Press, 2017.

Brockman, John, ed. *What to Think about Machines That Think: Today's Leading Thinkers on the Age of Machine Intelligence*. New York: Harper Perennial, 2015.

Broeckmann, Andreas. *Machine Art in the 20th Century*. Cambridge, MA: MIT Press, 2016.

Brooks, Rodney. "The Seven Deadly Sins of AI Predictions." *MIT Technology Review*, October 6, 2017, https://www.technologyreview.com/s/609048/the-seven-deadly-sins-of-ai-predictions.

Brooks, Rodney, and Anita Flynn. "Fast, Cheap, and Out of Control: A Robot Invasion of the Solar System." *Journal of the British Interplanetary Society* 42 (1989): 478–485. https://people.csail.mit.edu/brooks/papers/fast-cheap.pdf.

brown, adrienne maree. *Emergent Strategy: Shaping Change, Changing Worlds*. Chico, CA: AK Press, 2017.

Buolamwini, Joy, and Timnit Gebru. "Gender Shades: Intersectional Accuracy Disparities in Commercial Gender Classification." *Proceedings of Machine Learning Research*, 81:1–15. Conference on Fairness, Accountability, and Transparency, 2018. http://proceedings.mlr.press/v81/buolamwini18a/buolamwini18a.pdf.

Buolamwini, Joy. Written testimony to US Congress, May 22, 2019, https://www.congress.gov/116/meeting/house/109521/witnesses/HHRG-116-GO00-Wstate-BuolamwiniJ-20190522.pdf.

Campbell, Charlie. "Exclusive: The Chinese Scientist Who Sequenced the First COVID-19 Genome Speaks Out about the Controversies Surrounding His Work." *Time*, August 24, 2020, https://time.com/5882918/zhang-yongzhen-interview-china-coronavirus-genome.

"Case Study: Question Bridge Black Males." Media Impact Funders, January 2012, https://mediaimpactfunders.org/wp-content/uploads/2016/06/QuestionBridge.pdf.

Catlow, Ruth, March Garrett, Nathan Jones, and Sam Skinner, eds. *Artists Re:thinking the Blockchain*. Liverpool: Liverpool University Press, 2017.

Cecco, Leyland. "Google Affiliate Sidewalk Labs Abruptly Abandons Toronto Smart City Project." *The Guardian*, May 7, 2020, https://www.theguardian.com/technology/2020/may/07/google-sidewalk-labs-toronto-smart-city-abandoned.

Chang, Vanessa. *The Grid Art + Tech Report 2020*. EUNIC Silicon Valley, n.d., https://09227776-4c3f-45f5-9037-85b84a0e59af.filesusr.com/ugd/f31f46_8151ab52dbe040a6af559dc574be57d3.pdf?fbclid=IwAR2BTgInAbg4LKPLNAhVGvc7SG86P.

Chang, Vanessa. "Recoding the Master's Tools: Artists Remake Systems of Oppression and Extraction in Technology." National Endowment for the Arts, n.d., https://www.arts.gov/impact/media-arts/arts-technology-scan/essays/recoding-masters-tools-artists-remake-systems-oppression-and-extraction-technology.

Christian, Brian. *The Most Human Human*. New York: Anchor, 2012.

Cizek, Katerina, and Mandy Rose. "Documentary as a Co-Creative Practice: Interview with Kat Cizek." In *I-Docs: The Evolving Practices of Interactive Documentary*, ed. Judith Aston, Sandra Gaudenzi, and Mandy Rose, 38–48. New York: Columbia University Press, 2017.

Cizek, Katerina, and William Uricchio, with coauthors Juanita Anderson, Maria Aqui Carter, Detroit Narrative Agency, Thomas Allen Harris, Maori Karmael Holmes, Richard Lachman, Louis Massiah,

Cara Mertes, Sara Rafsky, Michèle Stephenson, Amelia Winger-Bearskin, and Sarah Wolozin. *Collective Wisdom: Co-Creating Media within Communities, across Disciplines and with Algorithms.* Cambridge, MA: MIT Press Works in Progress, 2019. https://wip.mitpress.mit.edu/collectivewisdom.

Coffman, Elizabeth. "Documentary and Collaboration: Placing the Camera in the Community." *Journal of Film and Video* 61, no. 1 (2009): 62–78.

"Collaborative Journalism Database." Center for Cooperative Media, https://collaborativejournalism.org/database-search-sort-learn-collaborative-projects-around-world.

"Community Benefits 101." The Parkdale People's Economy Project (blog), July 28, 2017, https://parkdalecommunityeconomies.wordpress.com/2017/07/28/community-benefits-101.

Cornock, Stroud, and Ernest Edmonds "The Creative Process Where the Artist Is Amplified or Superseded by the Computer." *Leonardo* 6, no. 1 (1973): 11–16.

Costanza-Chock, Sasha. "Design Justice, A.I., and Escape from the Matrix of Domination." *Journal of Design and Science,* July 16, 2018, https://jods.mitpress.mit.edu/pub/costanza-chock.

Costanza-Chock, Sasha. *Design Justice: Community-Led Practices to Build the Worlds We Need.* Cambridge, MA: MIT Press, 2020.

Costanza-Chock, Sasha. *Out of the Shadows, into the Streets! Transmedia Organizing and the Immigrant Rights Movement.* Cambridge, MA: MIT Press, 2014.

Council of the Haida Nation v. Province of British Columbia and Canada, No. L020662 (Vancouver, 2002), https://www.haidanation.ca/wp-content/uploads/2017/03/Statement_of_Claim.pdf.

"CPJ's Database on Attacks on the Press." The Committee to Protect Journalists, n.d., https://cpj.org/data/killed/?status=Killed&motiveConfirmed%5B%5D=Confirmed&type%5B%5D=Journalist&start_year=1992&end_year=2021&group_by=year.

Crabtree, Robin D. "Community Radio in Sandinista Nicaragua, 1979–1992: Participatory Communication and the Revolutionary Process." *Historical Journal of Film, Radio and Television* 16 no. 2 (1996): 221–241.

Crawford, Kate. *Atlas of AI: Power, Politics and the Planetary Costs of Artificial Intelligence.* New Haven, CT: Yale University Press, 2021.

Crawford, Kate, and Vladan Joler. "Anatomy of an AI System: The Amazon Echo as An Anatomical Map of Human Labor, Data and Planetary Resources." AI Now Institute and Share Lab, September 7, 2018, https://anatomyof.ai.

"Creative Futures." Ford Foundation, n.d., https://www.fordfoundation.org/campaigns/creative-futures/?popup=14896.

Cross, June. "We, the People." *Immerse*, July 26, 2019, https://immerse.news/we-the-people-f3947f23c792.

Cusi Wortham, Erica. *Indigenous Media in Mexico: Culture, Community, and the State.* Durham, NC.: Duke University Press, 2013.

Cutter, Laura. "Walter Reed, Yellow Fever, and Informed Consent." *Military Medicine* 181, no. 1 (2016): 90–91.

Czarnecki, Gina. "On Our Collaboration." Gina Czarnecki (blog), January 5, 2018, https://www .ginaczarnecki.com/single-post/2018/01/05/On-our-collaboration.

Darling, Kate. *The New Breed: What Our History with Animals Reveals about Our Future with Robots.* New York: Henry Holt, 2021.

De Carvalho, Ernesto Ignacio. "Video nas Aldeias: Frontiers." *In Media Res: A Media Commons Project Website,* May 8, 2009, http://mediacommons.org/imr/2009/05/08/v-deo-nas-aldeias-frontiers.

De Certeau, Michel. *The Practice of Everyday Life.* Berkeley: University of California Press, 1984.

Deibert, Ronald. *Reset: Reclaiming the Internet for Civil Society.* Toronto: Anansi, 2020.

DeJarnette, Ben. "Journalists and Academics Collaborating? It's Paying Off for Investigative Reports in Canada." *MediaShift* (blog), July 1, 2016, http://mediashift.org/2016/07/journalists -academics-collaborating-its-paying-off-investigative-reports-canada.

Deleuze, Gilles, and Felix Guattari. *A Thousand Plateaus: Capitalism and Schizophrenia.* Minneapolis: University of Minnesota Press, 1987.

Delwiche, Aaron, and Jennifer Jacobs Henderson, eds. *The Participatory Cultures Handbook.* New York: Routledge, 2013. http://www.gbv.de/dms/sub-hamburg/715341901.pdf.

De Michiel, Helen. "Documentary Untethered, Documentary Becoming." In Dossier Speculations and Inquiries on New Participatory Documentary Environments. *Afterimage: The Journal of Media Arts and Cultural Criticism* (2017).

De Michiel, Helen, and P. R. Zimmerman. "Documentary as an Open Space." In *The Documentary Film Book,* ed. B. Winston, 356–365. London: Palgrave Macmillan, 2013.

DePollo, Adam. "Riding with Aunt D. Dot: An Interview with Detroit Artist Bree Gant." *Michigan Quarterly Review,* n.d., https://sites.lsa.umich.edu/mqr/2018/03/riding-with-aunt-ddot-an-interview -with-detroit-artist-bree-gant.

"Design Justice Network Principles." Design Justice Network, Summer 2018, https://designjustice .org/read-the-principles.

DeVigal, Andrew. "The Continuum of Engagement: Pushing the Boundaries of News Engagement Day." *Let's Gather,* October 3, 2017, https://medium.com/lets-gather/the-continuum-of -engagement-89778f9d6c3a.

De Voe, Merrill. *How to Tailor Your Sales Organization to Your Markets.* New York: Prentice Hall, 1964.

Ding, Sue. "Interview with Hyphen-Labs." *Docubase,* n.d., https://docubase.mit.edu/lab/interviews /by-htm.

Dinkins, Stephanie. "Afro-now-ism." *Noēma,* June 16, 2020, https://www.noemamag.com/afro -now-ism.

Dinkins, Stephanie. "Conversations with Bina48." Stephanie Dinkins website, 2014–ongoing, https://www.stephaniedinkins.com/conversations-with-bina48.html.

Dixon, Steve. *Digital Performance: A History of New Media in Theater, Dance, Performance Art, and Installation.* Cambridge, MA: MIT Press, 2007.

Dovey, Jon. "Documentary Ecosystems: Collaboration and Exploitation." In *New Documentary Ecologies: Emerging Platforms, Practices and Discourses*, ed. K. Nash, C. Hight, and C. Summerhayes, 11–27. London: Palgrave Macmillan, 2014.

Dovey, Jon, and Mandy Rose. "'This Great Mapping of Ourselves': New Documentary Forms Online." In *The Documentary Film Book*, ed. B. Winston, 366–375. London: Palgrave Macmillan, 2013. http://eprints.uwe.ac.uk/17085/2/BFIDovey-Rose2012.pdf.

Durham Peters, John. *The Marvelous Clouds: Toward a Philosophy of Elemental Media*. Chicago: University of Chicago Press, 2015.

Eco, Umberto. *The Open Work*. Cambridge, MA: Harvard University Press, 1989. https://monoskop .org/images/6/6b/Eco_Umberto_The_Open_Work.pdf.

Edmonds, Ernest. *The Art of Interaction: What HCI Can Learn from Interactive Art*. Pittsburgh: Morgan and Claypool Publishers, 2018.

Ehn, Pelle, Elisabet M. Nilsson, and Richard Topgaard, eds. *Making Futures: Marginal Notes on Innovation, Design, and Democracy*. Cambridge, MA: Harvard University Press, 2014.

Elder, Sarah. "Collaborative Filmmaking: An Open Space for Making Meaning, a Moral Ground for Ethnographic Film." *Visual Anthropology Review* 11, no. 2 (1995): 94–101. Enwezor, Okwui. "The Author as Producer in Times of Crisis." *Dark Matter,* 2004, http://www.darkmatterarchives .net/wp-content/uploads/2011/02/Enwezor.AuthorProd.pdf.

Escobar, Arturo. *Designs for the Pluriverse*. Durham, NC: Duke University Press, 2017.

Etim, Bassey. "The Times Sharply Increases Articles Open for Comments, Using Google's Technology." *New York Times*, June 13, 2017. https://www.nytimes.com/2017/06/13/insider/have-a-comment -leave-a-comment.html.

Evans, Mike, and Stephen Foster. "Representation in Participatory Video: Some Considerations from Research with Métis in British Columbia." *Journal of Canadian Studies/Revue d'études canadiennes* 43(1) (2009): 87–108.

Fidler, Courtney, and Michael Hitch. "Impact and Benefit Agreements: A Contentious Issue for Environmental and Aboriginal Justice." *Environments Journal* 35(2) (2007): 49–69. https://journals .scholarsportal.info/pdf/07116780/v35i0002/2_iabaacifeaaj.xml.

Finn, Ed. *What Algorithms Want: Imagination in the Age of Computing*. Cambridge, MA: MIT Press, 2017.

Fitzgibbon, Will. "What Happened after the Panama Papers?" *International Consortium of Investigative Journalists*, April 3, 2019, https://www.icij.org/investigations/panama-papers/what-happened -after-the-panama-papers/.

Flores, Carlos Y. "Indigenous Video, Development and Shared Anthropology: A Collaborative Experience with Maya Q'eqchi Filmmakers in Postwar Guatemala." *Visual Anthropology Review* 20, no. 1 (2004): 31–44.

Foucault, Michel. "What Is an Author?" In *Textual Strategies: Perspectives in Post-Structuralist Criticism*, ed. Josue V. Harari, 141–160. Ithaca, NY: Cornell University Press, 1969. http://www .generation-online.org/p/fp_foucault12.htm.

Frankham, Bettina. "Moving beyond Evidence: Participatory Online Documentary Practice within the Poetic Framework of Cowbird." *Media International Australia* 154, no. 1 (2015): 123–131.

Freelon, Deen, Lori Lopez, Meredith D. Clark, and Sarah J. Jackson. "How Black Twitter and Other Social Media Communities Interact with Mainstream News." Knight Foundation, 2017, https://knightfoundation.org/features/twittermedia.

Freeman, Jo. "The Tyranny of Structurelessness." *Second Wave* 2, no. 1 (1972): 20.

Freire, Paulo. *Pedagogy of the Oppressed*. New York: Herder and Herder, 1970.

Galloway, Alex, and Eugene Thacker. *The Exploit: A Theory of Networks*. Minneapolis: University of Minnesota Press, 2007. http://dss-edit.com/plu/Galloway-Thacker_The_Exploit_2007.pdf.

Gaudenzi, Sandra. "The Living Documentary: From Representing Reality to Co-creating Reality in Digital Interactive Documentary." Doctoral thesis, Goldsmiths, University of London, 2013.

Gaudenzi, Sandra. "Strategies of Participation: The Who, What and When of Collaborative Documentaries." In *New Documentary Ecologies: Emerging Platforms, Practices and Discourses*, ed. K. Nash, C. Hight, and C. Summerhayes, 129–148. London: Palgrave Macmillan, 2014.

Gebeloff, Robert, Danielle Ivory, Matt Richtel, Mitch Smith, Karen Yourish, Jackie Fortiér, Elly Yu, and Molly Parker. "The Striking Racial Divide in How Covid-19 Has Hit Nursing Homes." *The New York Times*, September 10, 2020, https://www.nytimes.com/article/coronavirus-nursing-homes-racial-disparity.html.

George, Babitha. "Making Space for the Informal: Interview with Rainman." Medium, March 9, 2016, https://medium.com/reflecting-on-toolkits/making-space-for-the-informal-99f0968d1bc0.

Ghosh, Rishab. *CODE: Collaborative Ownership and the Digital Economy*. Cambridge, MA: MIT Press, 2006.

Goldstein, Bruce E. *Collaborative Resilience*. Cambridge, MA: MIT Press, 2011.

Graeber, David. *Debt: The First 5000 Years*. New York: Melville House, 2011.

Graham, Zoe. "'Since You Are Filming, I Will Tell the Truth': A Reflection on the Cultural Activism and Collaborative Filmmaking of Video Nas Aldeias." *Visual Anthropology Review* 30 no. 1 (2014): 89–91.

Gregory, Sam. "It Gets Better: Collective and Individual Voice in Video Advocacy." WITNESS (blog), n.d., https://blog.witness.org/2010/11/it-gets-better-collective-and-individual-voice-in-video-advocacy/.

Gubrium, Aline, and Krista Harper. *Participatory Visual and Digital Methods*. New York: Routledge, 2016.

Gye, Lisa. "Docummunity and the Disruptive Potential of Collaborative Documentary Filmmaking." *Expanding Documentary 2011: Proceedings of the VIIIth Biennial Conference*, ed. G. Peters. Auckland, NZ: Auckland University of Technology, the School of Communication Studies, Faculty of Design & Creative Technologies, 2011. https://researchbank.swinburne.edu.au/file/9810bd5c-82f7-4dfc-8fcd-ce9de32ff2c3/1/PDF%20%28Published%20version%29.pdf.

Haff, Peter. "Humans and Technology in the Anthropocene: Six Rules." *Anthropocene Review* 1, no. 2 (2014): 126–136.

Hansen, Mark, Meritxell Roca-Sales, Jonathan M. Keegan, and George King. "Artificial Intelligence: Practice and Implications for Journalism." *Columbia University Libraries, Academic Commons,* 2017, https://academiccommons.columbia.edu/catalog/ac:gf1vhhmgs8.

Hao, Karen. "How Facebook Got Addicted to Spreading Misinformation," *MIT Technology Review,* March 11, 2021, https://www.technologyreview.com/2021/03/11/1020600/facebook-responsible -ai-misinformation/.

Harari, Yuval Noah. *Homo Deus: A Brief History of Tomorrow.* New York: HarperCollins, 2017.

Haraway, Donna. *The Companion Species Manifesto: Dogs, People, and Significant Otherness.* Chicago: Prickly Paradigm, 2003.

Hardt, Michael, and Antonio Negri. *Multitude: War and Democracy in the Age of Empire.* New York: Penguin Books, 2004.

Hardy, Forsyth, ed. *Grierson on Documentary.* Berkeley: University of California Press, 1971.

Harvey, Kerric. "'Walk-In Documentary': New Paradigms for Game-Based Interactive Storytelling and Experiential Conflict Mediation." *Studies in Documentary Film* 6, no. 2 (2014): 189–202.

Heraclitus of Ephesus. *Heraclitus on the Two Antithetical Forces in Life.* Trans. G. W. T. Patrick. London: Philaletheians, 2017.

Hern, Alex. "New AI Fake Text Generator May Be Too Dangerous to Release, Say Creators." *The Guardian,* February 14, 2019, https://www.theguardian.com/technology/2019/feb/14/elon-musk -backed-ai-writes-convincing-news-fiction?fbclid=IwAR3IEShRs0kUW8-h29qCunefcMz62CQOhz FYKzx7oSVnHh4NaZWkYvhohrY.

"History of the Haida Nation." Haida Nation, n.d., https://www.haidanation.ca/?page_id=26.

Hoechner, Hannah. "Participatory Filmmaking with Qur'anic Students in Kano, Nigeria: 'Speak Good about Us or Keep Quiet!'" *International Journal of Social Research Methodology* 18, no. 6 (2013): 635–649.

Hofstadter, Douglas. *Gödel, Escher, Bach: An Eternal Golden Braid.* New York: Basic Books, 1979.

Houser, Kristin. "AI-Generated Art Will Go on Sale alongside Human-Made Works This Fall," *Futurism,* August 22, 2018, https://futurism.com/the-byte/ai-generated-portrait-christies.

Hyphen-Labs. n.d., http://www.hyphen-labs.com.

Isaacson, Walter. *The Code Breaker: Jennifer Doudna, Gene Editing, and the Future of the Human Race.* New York: Simon & Schuster, 2021.

Isaacson, Walter. "CRISPR Rivals Put Patents aside to Help in Fight against Covid-19." *STAT,* March 3, 2021, https://www.statnews.com/2021/03/03/crispr-rivals-put-patents-aside-fight-against -covid-19/?utm_source=STAT+Newsletters&utm_campaign=caa583f0de-MR_COPY_14&utm _medium=email&utm_term=0_8cab1d7961-caa583f0de-152446341.

Ito, Joichi. "Antidisciplinary." Joi Ito (blog). Accessed March 23, 2021. https://joi.ito.com/weblog/2014/10/02/antidisciplinar.html.

Ito, Joichi. "Resisting Reduction: A Manifesto." *Journal of Design and Science*, October 13, 2017. https://jods.mitpress.mit.edu/pub/resisting-reduction/release/17.

Jackson, Tabitha. "The Art of Co-Creation: A Storytelling Model for Impact and Engagement." YouTube, April 13, 2018, https://www.youtube.com/watch?v=ALCwYtD2x3E&ab_channel=Skoll.org.

Jahanshahi, Mina. "The Greatest Land Art Project in Iran." The Iranian, April 15, 2018, https://iranian.com/2018/04/15/greatest-land-art-project-iran/.

Jahn, Marissa, ed. *ByProduct: On the Excess of Embedded Art Practices*. Toronto: YYZ, 2010.

Janke, Terri. *Pathways and Protocols: A Filmmaker's Guide to Working with Indigenous People, Culture and Concepts*. n.p.: Australian Government, Screen Australia, 2009. https://www.screenaustralia.gov.au/getmedia/16e5ade3-bbca-4db2-a433-94bcd4c45434/Pathways-and-Protocols/.

Jeffries, Stuart. "When Two Tribes Meet: Collaborations between Artists and Scientists." *The Guardian*, August 21, 2011, https://www.theguardian.com/artanddesign/2011/aug/21/collaborations-between-artists-and-scientists.

Jenkins, Henry. *Confronting the Challenges of Participatory Culture: Media Education for the 21st Century*. Cambridge, MA: MIT Press, 2009.

Jenkins, Henry. "Digital Land Grab." *MIT Technology Review* (blog), March 1, 2000, https://www.technologyreview.com/s/400696/digital-land-grab.

Jenkins, Henry. *Fans, Bloggers, and Gamers: Exploring Participatory Culture*. New York: New York University Press, 2006.

Jenkins, Henry. *Textual Poachers: Television Fans and Participatory Culture*. New York: Routledge, 1992.

Jenkins, Henry, and N. Carpentier. "Theorizing Participatory Intensities: A Conversation about Participation and Politics." *Convergence: The International Journal of Research into New Media Technologies* 19, no. 3 (2013): 265–286.

Jenkins, Henry, Sam Ford, and Joshua Green. *Spreadable Media: Creating Value and Meaning in a Networked Culture*. New York: New York University Press, 2013.

Jenkins, Henry, Sangita Shresthova, Liana Gamber Thompson, Neta Kligler-Vilenchik, and Arely M. Zimmerman. *By Any Media Necessary: The New Youth Activism*. New York: New York University Press, 2016.

Johnson, Joseph. "Global Digital Population as of January 2021." *Statista*, March 5, 2021, https://www.statista.com/statistics/617136/digital-population-worldwide/#:~:text=As%20of%20January%202021%20there,the%20internet%20via%20mobile%20devices.

Johnson, Steven. *Emergence: The Connected Lives of Ants, Brains, Cities, and Software*. New York: Scribner, 2002.

Jolly, Jihii. "Platform Aimed at Audience Interaction Generates Story Ideas, Goodwill." *Columbia Journalism Review*, August 22, 2016, https://www.cjr.org/the_profile/hearken_hey_area_homeless_san_francisco_audience.php?utm_content=bufferc4c9f&utm_medium=social&utm_source=twitter.com&utm_campaign=buffer.

Joseph, George, and Kenneth Lipp. "IBM Used NYPD Surveillance Footage to Develop Technology That Lets Police Search by Skin Color." The Intercept, September 6, 2018. https://theintercept.com/2018/09/06/nypd-surveillance-camera-skin-tone-search.

"Journalists, Academics Discuss Collaboration in 'Mind to Mind' Symposium." *Reveal* (blog), October 23, 2017, https://revealnews.org/press/journalists-academics-discuss-collaboration-in-mind-to-mind-symposium.

Juhasz, Alexandra. "No Woman Is an Object: Realizing the Feminist Collaborative Video." *Camera Obscura* 18, vol. 2 (2003): 70–97.

Juhasz, Alexandra, and Alisa Lebow, eds. "Beyond Story: Ways to Engage." *World Records* 5 (2021), https://vols.worldrecordsjournal.org/05/01?index=1.

Kafai, Yasmin B., and Quinn Burke. *Connected Gaming: What Making Video Games Can Teach Us about Learning and Literacy*. Cambridge, MA: MIT Press, 2016.

Kandell, Eric. *The Age of Insight: The Quest to Understand the Unconscious in Art, Mind, and Brain, from Vienna 1900 to the Present*. New York: Random House, 2012.

Kapur, Anandana. "Using the Collaborative and Interactive Docu-forms to (Re)imagine the 'Rape City.'" In *I-docs: The Evolving Practices of Interactive Documentary*, ed. Judith Aston, Sandra Gaudenzi, and Mandy Rose, 26–37. New York: Columbia University Press, 2017.

Keating, Joshua. "The George Floyd Protests Show Leaderless Movements Are the Future of Politics." *Slate*, June 9, 2020, https://slate.com/news-and-politics/2020/06/george-floyd-global-leaderless-movements.html.

Kelly, Kevin. *Out of Control: The New Biology of Machines, Social Systems, and the Economic World*. New York: Basic Books, 1995.

Kember, Sarah, and Joanna Zylinska. *Life after New Media: Mediation as a Vital Process*. Cambridge, MA: MIT Press, 2012.

Kerrigan, Susan. "Collaborative and Creative Documentary Production in Video and Online." In *Proceedings of ISEA2008: The 14th International Symposium on Electronic Art*, 265–267. Singapore: ISEA, 2008. http://www.isea-archives.org/docs/2008/proceedings/ISEA2008_proceedings.pdf.

Kester, Grant H. *The One and the Many: Contemporary Collaborative Art in a Global Context*. Durham, NC: Duke University Press, 2011.

Khan, Aiman, and Ishita Chakrabarty, "Why the 2020 Violence in Delhi Was a Pogrom." Al Jazeera, February 24, 2021, https://www.aljazeera.com/opinions/2021/2/24/why-the-2020-violence-in-delhi-was-a-pogrom.

Khan, Nora. "Fred Turner: Silicon Valley Thinks Politics Doesn't Exist." *032c*, July 30, 2018, https://032c.com/fred-turner-silicon-valley-thinks-politics-doesnt-exist.

Kilborn, Richard, Matthew Hibberd, and John Izod, eds. *From Grierson to the Docu-Soap: Breaking the Boundaries*. Luton, UK: University of Luton Press, 2000.

Kim, Angie. "The Co-Op." Ford Foundation, n.d., https://www.fordfoundation.org/campaigns/creative-futures/?popup=angie-kim.

Kindon, Sara. "Participatory Video in Geographic Research: A Feminist Practice of Looking?" *Area* 35, no. 2 (2003): 142–153.

Kinsella, Hank Eileen. "Willis Thomas Pulls Work from a South African Art Fair after a Photographer Levels Plagiarism Charges," Artnet, September 13, 2018, https://news.artnet.com/art-world/artist-hank-willis-thomas-pulls-work-from-art-fair-after-photographer-levels-plagiarism-charges-1347710.

konaté, nènè myriam, and Lucas LaRochelle. "Glitch as Wormhole." *Immerse,* October 29, 2020, https://immerse.news/glitch-as-wormhole-a400e4af905.

Kummels, Ingrid. "Ser comunero en tiempos de diáspora: Videos de fiesta en el distrito de Mixe, Oaxaca, México." In *Dentro y fuera de cuadro: Identidad, representación y auto representación visual de los pueblos indígenas de América Latina, Siglos XIX y XX*, ed. Margarita Alvarado, and María Paz Bajas. Santiago, Chile: Pehuén Editores, 2014.

Lam, Thuy Vo. "Adversarial Experiments Everywhere." *Creative Futures*, March 14, 2021, https://www.fordfoundation.org/campaigns/creative-futures/?popup=14999.

LaSalle, Daniel. *Summary and Index of Community Benefit Agreements*. New Orleans: Public Law Center, 2011. https://law.tulane.edu/sites/law.tulane.edu/files/Files/TPLC/summary-and-index-community-benefit-agreements.pdf.

Lecco, Leyland. "Google Affiliate Sidewalk Labs Abruptly Abandons Toronto Smart City Project." *The Guardian*, May 7, 2020, https://www.theguardian.com/technology/2020/may/07/google-sidewalk-labs-toronto-smart-city-abandoned.

Levitan, Tyler, and Emilie Cameron. "Privatizing Consent: Impact and Benefit Agreements and the Neoliberalization of Mineral Development in the Canadian North." In *Mining and Communities in Northern Canada: History, Politics, and Memory*, ed. Arn Keeling and John Sandlos. Calgary, Canada: University of Calgary Press, 2015.

Levy, David. *Love and Sex with Robots: The Evolution of Human-Robot Relationships*. New York: HarperCollins, 2007.

Levy, Pierre. *Collective Intelligence: Mankind's Emerging World in Cyberspace*. New York: Basic Books, 1999.

Lewis, Jason Edward, Noelani Arista, Archer Pechawis, and Suzanne Kite. "Making Kin with the Machines." *Journal of Design and Science*, July 16, 2018. https://doi.org/10.21428/bfafd97b.

Literat, Ioana. "The Work of Art in the Age of Mediated Participation: Crowdsourced Art and Collective Creativity." *International Journal of Communication* 6 (2012): 2962–2984.

Lohr, Steve. "M.I.T. Plans College for Artificial Intelligence, Backed by $1 Billion." *New York Times*, October 15, 2018, https://www.nytimes.com/2018/10/15/technology/mit-college-artificial-intelligence.html.

Lorberbaum, Tal, Kevin J. Sampson, Raymond L. Woosley, Robert S. Kass, and Nicholas P. Tatonetti. "An Integrative Data Science Pipeline to Identify Novel Drug Interactions That Prolong the QT Interval." *Drug Safety* 39 (2016): 433–441. https://www.ncbi.nlm.nih.gov/pmc/articles/PMC4835515/pdf/40264_2016_Article_393.pdf.

Low, Colin. "Grierson and Challenge for Change (1984)." In *Challenge for Change: Activist Documentary at the National Film Board of Canada*, ed. Thomas Waugh, Michael Brendan Baker, and Ezra Winton, 16–23. Montreal: McGill-Queen's University Press, 2010.

Lowenthal, Andrew, Tanya Notley, Lina Srivastava, and Egbert Wits. "Video for Change Impact Toolkit." Video for Change, 2019, https://toolkit.video4change.org.

Löwgren, Jonas, and Bo Reimer. *Collaborative Media: Production, Consumption, and Design Interventions*. Cambridge, MA: MIT Press, 2013.

Lui, Bei, Jianlong Fu, Makoto P. Kato, and Masatoshi Yoshikawa. "Beyond Narrative Description: Generating Poetry from Images by Multi-Adversarial Training." 2018 ACM Multimedia Conference (MM '18), October 22–26, 2018, Seoul, Republic of Korea. https://doi.org/10.1145/3240508 .3240587.

Lunch, Nick, and Chris Lunch. *Insights into Participatory Video: A Handbook for the Field*. Oxford, UK: InsightShare, 2006.

Lynas, Mark. *The God Species: Saving the Planet in the Age of Humans*. Washington, DC: National Geographic Society, 2011.

Marcello, David A. "Community Benefit Agreements: New Vehicle for Investment in America's Neighborhoods." *Urban Lawyer* 39, no. 3 (2007): 657–669.

Martin, C. Dianne. "ENIAC: The Press Conference That Shook the World." *IEEE Technology and Society Magazine*, December 1995.

Martin, Rachel L. "Making *Eyes on the Prize: An Oral History*." Ford Foundation, n.d., https://www .fordfoundation.org/just-matters/ford-forum/making-eyes-on-the-prize-an-oral-history/.

Martínez de la Serna, Carlos. "Collaboration and the Creation of a New Journalism Commons." *Columbia Journalism Review*, March 30, 2018, https://www.cjr.org/tow_center_reports/collaboration -and-the-journalism-commons.php.

Marx, Tony, and Jamie Woodson. *Crisis in Democracy: Renewing Trust in America*. Washington, DC: Aspen Institute/Knight Foundation, 2019. https://csreports.aspeninstitute.org/documents /Knight2019.pdf.

Mayer, Joy. "It's Time to Gather, and We Hope You'll Join Us." *Let's Gather*, October 1, 2017, https://medium.com/lets-gather/its-time-to-gather-and-we-hope-you-ll-join-us-85700c4a9fe3.

McCall, Sophie. *First Person Plural: Aboriginal Storytelling and the Ethics of Collaborative Authorship*. Vancouver: University of British Columbia Press, 1969.

McGuire, Mark Patrick. "Trading a Notebook for a Camera: Toward a Theory of Collaborative, Ethnopoetic Filmmaking." *Contemporary Buddhism: An Interdisciplinary Journal* 15 (2014): 164–198.

McKenzie, Scott. *Film Manifestos and Global Cinema Cultures*. Berkeley: University of California Press, 2014.

McLuhan, Marshall, and Quentin Fiore. *The Medium Is the Massage: An Inventory of Effects*. New York: Bantam Books, 1967.

McLuhan, Marshall, and Michael Moos. *Media Research: Technology, Art and Communication*. New York: Routledge, 1997.

McNally, Kevin. "It's Time to Stop Using AI as a Marketing Gimmick." *Fast Company*, July 18, 2017, https://www.fastcompany.com/90133598/its-time-to-stop-using-ai-as-a-marketing-gimmick.

Merrill, De Voe. *How to Tailor Your Sales Organization to Your Markets*. New York: Prentice Hall, 1964.

Miller, Elizabeth. *Going Public: The Art of Participatory Practice*. Vancouver: University of British Columbia Press, 2017.

Milne, E-J, Claudia Mitchell, and Naydene de Lange, eds. *Handbook of Participatory Video*. Plymouth, UK: AltaMira, 2012.

Mindell, David. *Our Robots, Ourselves: Robotics and the Myths of Autonomy*. New York: Viking, 2015.

Minksy, Marvin. *The Emotion Machine: Commonsense Thinking, Artificial Intelligence, and the Future of the Human Mind*. New York: Simon & Schuster, 2007.

Minksy, Marvin. *The Society of Mind*. New York: Simon & Schuster, 1988.

Mirani, Leo. "Millions of Facebook Users Have No Idea They're Using the Internet." *Quartz*, February 9, 2015, https://qz.com/333313/milliions-of-facebook-users-have-no-idea-theyre-using-the-internet.

Mitchell, Mary. "Where Voice and Listening Meet: Participation in and through Interactive Documentary in Peru." *Glocal Times: The Communication for Development Journal* 22/23 (2015): 1–13.

Mori, Masahiro. *The Buddha in the Robot*. Tokyo: Kosei Publishing Company, 1989.

Morozov, Evgeny. *To Save Everything, Click Here: The Folly of Technological Solutionism*. New York: PublicAffairs, 2013.

Mozur, Paul. "A Genocide Incited on Facebook, with Posts from Myanmar's Military." *The New York Times*, October 15, 2018, https://www.nytimes.com/2018/10/15/technology/myanmar-facebook-genocide.html.

MSL Group. "Co-creation Communities—Ten Frontiers for the Future of Engagement." *MSL People's Insights Annual Report* (2013). https://issuu.com/mslgroupofficial/docs/5-co-creation-communities-future-engagement/6.

Nash, Kate. *New Documentary Ecologies: Emerging Platforms, Practices and Discourses*. London: Palgrave Macmillan, 2014.

Nash, Kate. "What Is Interactivity For? The Social Dimension of Web-Documentary Participation." *Continuum: Journal of Media & Cultural Studies* 28, no. 3 (2014): 383–395.

Nemtin, Bill, and Colin Low. *Fogo Island Film and Community Development Project*. Ottawa: National Film Board of Canada, 1968. http://onf-nfb.gc.ca/medias/download/documents/pdf/1968-Fogo-Island-Project-Low-Nemtin.pdf.

"New Frontiers in Research Fund." Government of Canada, November 27, 2020, https://www.sshrc-crsh.gc.ca/funding-financement/nfrf-fnfr/index-eng.aspx.

"New Navy Device Learns by Doing; Psychologist Shows Embryo of Computer Designed to Read and Grow Wiser." *New York Times*, July 8, 1958. 25.

Nickerson, Marcia. *On Screen Protocols & Pathways: A Media Production Guide to Working with First Nations, Métis and Inuit Communities, Cultures, Concepts, and Stories*. Toronto: imagiNative, 2019.

Noble, Safiya. *Algorithms of Oppression*. New York: New York University, 2018.

Norman, Donald A. *The Design of the Everyday Things*. New York: Basic Books, 2002.

Norman, Donald A., and Stephen W. Draper, eds. *User Centered System Design: New Perspectives on Human-Computer Design*. Hillsdale, NJ: Lawrence Erlbaum Associates, 1986.

Notley, Tanya, Sam Gregory, and Andrew Lowenthal. "Video for Change: Creating and Measuring Ethical Impact." *Journal of Human Rights Practice* 9, no. 2 (2017): 223–246.

"NSAF." Hyphen-Labs, n.d., http://www.hyphen-labs.com/nsaf.html.

O'Flynn, Siobhan. "Documentary's Metamorphic Form: Webdoc, Interactive, Transmedia, Participatory and Beyond." *Studies in Documentary Film* 6, no. 2 (2014): 141–157.

Olariu, Stephan, and Albert Y. Zomaya. *Handbook of Bioinspired Algorithms and Applications*. Boca Raton, FL: CRC, 2005.

Ornuoha, Mimi (*Mother Cyborg*). *A People's Guide to Artificial Intelligence*. Detroit: Whitlock, 2018. https://www.alliedmedia.org/files/peoples-guide-ai.pdf.

Pack, Sam. "Collaborative Filmmaking in the Digital Age." *Anthropology Now* 4, no. 1 (2021): 85–89.

Parikka, Jussi. *Insect Media: An Archaeology of Animals and Technology*. Minneapolis: University of Minnesota Press, 2010.

Parker, Priya. *The Art of Gathering: How We Meet and Why It Matters*. New York: Riverhead Books, 2018.

Parkin, Simon. "The YouTube Stars Heading for Burnout: 'The Most Fun Job Imaginable Became Deeply Bleak.'" *The Guardian*, September 8, 2018, https://www.theguardian.com/technology/2018/sep/08/youtube-stars-burnout-fun-bleak-stressed?CMP=Share_iOSApp_Other.

Pateman, Carole. *Participation and Democratic Theory*. Cambridge: Cambridge University Press, 1970.

"Platform Co-op Resource Library." *Platform Cooperativism Consortium*, n.d., https://resources.platform.coop.

Porter, Catherine. "Reviving a Lost Language of Canada through Film," *New York Times*, June 11, 2017, https://www.nytimes.com/2017/06/11/world/americas/reviving-a-lost-language-of-canada-through-film.html.

Prahalad, C. K., and Venkat Ramaswamy. *The Future of Competition: Co-Creating Unique Value with Customers*. Cambridge, MA: Harvard University Press, 2004.

"Prepare, Don't Panic: Synthetic Media and Deepfakes." WITNESS *Media Labs* (blog), n.d., https://lab.witness.org/projects/synthetic-media-and-deep-fakes.

Project Facet. http://www.projectfacet.org.

"Proyecto Quipu: El drama de las esterilizaciones forzadas en la dictadura de Fujimori." *The Clinic Online*, February 28, 2016, https://www.theclinic.cl/2016/02/28/proyecto-quipu-el-drama-de-las -esterilizaciones-forzadas-en-la-dictadura-de-fujimori.

Public VR Lab. https://publicvrlab.squarespace.com.

"The Quipu Project: A Framework for Participatory Interactive Documentary." *I-docs,* n.d., http://i -docs.org/the-quipu-project-a-framework-for-participatory-interactive-documentary.

Raita, Yurei. *The Day a Computer Writes a Novel*. Sato-Matsuzaki Laboratory, 2016.

Rao, Sameer. "How 'Neurospeculative AfroFeminism' Uses Virtual Reality to Explore Other-worldly Transformation at a Black Hair Salon." *Colorlines*, March 17, 2017, https://www.colorlines .com/articles/how-neurospeculative-afrofeminism-uses-virtual-reality-explore-otherworldly -transformation.

Rangan, Pooja. *Immediations: The Humanitarian Impulse in Documentary*. Durham, NC: Duke University Press, 2017.

Ratti, Carlo, and Matthew Claudel. *Open Source Architecture*. London: Thames & Hudson, 2015.

"Reach.love." *Docubase,* n.d., https://docubase.mit.edu/tools/reach.

"React Report 2012–2016." *Research & Enterprise in Arts and Creative Technology* (REACT), 2012–2016, http://www.react-hub.org.uk/sites/default/files/publications/REACT%20Report%20low%20res_2 .pdf.

Redbird, Duke. *Dish with One Spoon*. Video filmed November 10, 2020, Open Documentary Lab Lecture Series, http://opendoclab.mit.edu/presents/duke-redbird-indigenous-digital-delegation.

Reynolds, Craig. "Flocks, Herds, and Schools: A Distributed Behavioral Model." *ACM SIGGRAPH Computer Graphics* 21, no. 4 (1987): 25–34.

Riddle, Anthony, and George Stoney. "Local Public Access TV under Attack from Trio of Congressional Bills." *Democracy Now!,* September 30, 2005, https://www.democracynow.org/2005/9/30 /local_public_access_tv_under_attack.

Riecken, Ed, Frank Conibear, Corrine Michel, John Lyall, Tish Scott, Michele Tanaka, Suzanne Stewart, Janet Riecken, and Teresa Strong-Wilson. "Resistance through Re-Presenting Culture: Aboriginal Student Filmmakers and a Participatory Action Research Project on Health and Wellness." *Canadian Journal of Education* 29, no. 1 (2006): 265–286.

Roe, Sam. "Could Collaborating with Scientists Be the Next Step for Investigative Reporting?" *Columbia Journalism Review*, February 12, 2016. https://www.cjr.org/first_person/could_collabo rating_with_scientists_be_the_next_step_for_investigative_reporting.php.

Rose, Mandy. "Making Publics: Documentary as Do-It-with-Others Citizenship." In *DIY Citizenship: Critical Making and Social Media*, ed. Matt Ratto and Megan Boler, 201–212. Cambridge, MA: MIT Press, 2014.

Rose, Mandy. "Not Media About, But Media With." In *I-docs: The Evolving Practices of Interactive Documentary*, ed. Judith Aston, Sandra Gaudenzi, and Mandy Rose, 49–65. New York: Columbia University Press, 2017.

Rose, Mandy. "Through the Eyes of the Video Nation." In *From Grierson to the Docu-Soap: Breaking the Boundaries*, ed. Richard Kilborn, Matthew Hibberd, and John Izod, 173–184. Luton, UK: University of Luton Press, 2000.

Rosner, Helen. "A Haunting New Documentary about Anthony Bourdain." *New Yorker*, July 15, 2021, https://www.newyorker.com/culture/annals-of-gastronomy/the-haunting-afterlife-of-anthony -bourdain.

Rowe, G. Peter. *Design Thinking*. Cambridge, MA: MIT Press, 1986.

Ruby, Jay. "Speaking For, Speaking About, Speaking With, or Speaking Alongside: An Anthropological and Documentary Dilemma." *Journal of Film and Video* 44, no. 1–2 (1992): 42–66.

Sample, Ian. "The Great Project: How Covid Changed Science Forever." *The Guardian*, December 15, 2020, https://www.theguardian.com/world/2020/dec/15/the-great-project-how-covid-changed -science-for-ever.

Sandercok, Leonie, Dana Moraes and Jonathan Frantz. "Film as a Catalyst for Indigenous Community Development." *Planning Theory and Practice* 18, no. 4 (2017): 639–666.

Sanders, Elizabeth B. N., and Pieter Jan Stappers. "Co-creation and the New Landscapes of Design." *Co-Design* 4, no. 1 (2008): 5–18.

Sanjinés, Jorge, and Scott MacKenzie. *Film Manifestos and Global Cinema Cultures: A Critical Anthology Problems of Form and Content in Revolutionary Cinema*. Berkeley: University of California Press, 2014.

Santos Lopes de Aguiar, Laura. *We Were There: The Women of the Maze and Long Kesh Prison: Collaborative Filmmaking in Transitional Northern Ireland*. Belfast, UK: Queen's University Belfast, 2017. http://ethos.bl.uk/OrderDetails.do?uin=uk.bl.ethos.705633.

SecureDrop. https://securedrop.org.

Schiwy, Freya, and Byrt Wammack Weber, eds. *Adjusting the Lens: Community and Collaborative Video in Mexico*. Pittsburgh, PA: University of Pittsburgh Press, 2017.

Scholtz, Trebor. *Uberworked and Underpaid: How Workers Are Disrupting the Digital Economy*. Cambridge, UK: Polity, 2016.

Scholz, Trebor, and Nathan Schneider. *Our to Hack and Own: The Rise of Platform Cooperativism; A New Vision for the Future of Work and a Fairer Internet*. New York: O/R Books, 2016.

Schudson, Michael. *Discovering the News: A Social History of American Newspapers*. New York: Dover Books, 1978.

Schwab, Klaus. *The Fourth Industrial Revolution*. Geneva, Switzerland: World Economic Forum, 2016.

Sinclair, Kamal. "Design Public Space." Medium, June 1, 2018, https://makinganewreality.org /design-public-space-228ab32965be.

Sinclair, Kamal. "The High Stakes of Limited Inclusion." Medium, November 29, 2017, https://makinganewreality.org/the-high-stakes-of-limited-inclusion-8c11541829f6.

Sinclair, Kamal. *"Making a New Reality: A Toolkit for Inclusive Media Futures."* Medium, (2018, makinganewreality.org.

Sinclair, Kamal. "Share Space." Medium, March 18, 2018, March 25, 2021. https://makinganewreality.org/share-space-fc4ff7ae6d5e.

Sitrin, Marina. "Definitions of Horizontalism and Autonomy." *NACLA: Horizontalism & Autonomy* 47, no. 3: (2014). https://nacla.org/article/definitions-horizontalism-and-autonomy.

SITU Research. "Euromaidan Event Reconstruction." Carnegie Mellon University, Center for Human Rights, 2018, https://situ.nyc/research/projects/euromaidan-event-reconstruction.

Slade, Samantha. *Going Horizontal: Creating a Non-Hierarchical Organization, One Practice at a Time.* Oakland, CA: Berrett-Koehler Publishers, 2018.

Smith, Linda Tuhiwai. *Decolonizing Methodologies: Research and Indigenous Peoples.* London: Zed Books, 1999. https://nycstandswithstandingrock.files.wordpress.com/2016/10/linda-tuhiwai-smith-decolonizing-methodologies-research-and-indigenous-peoples.pdf.

Soja, E. W. *Thirdspace: Journeys to Los Angeles and Other Real-and-Imagined Places.* Oxford, UK: Blackwell, 1996.

Soundararajan, Thenmozhi. *Socially Engaged Art in Volatile Times.* Video filmed February 16, 2021, at the Open Documentary Lab Lecture Series, Cambridge, MA. http://opendoclab.mit.edu/presents/thenmozhi-soundararajan-equality-labs.

Srivastava, Lina. "Born into Brothels, Five Years Later." Lina Srivastava: Narrative Strategies to Catalyze Transformational Change, December 2009, https://linasrivastava.blogspot.com/2009/12/born-into-brothels-five-years-later.html.

Stallman, Richard. *Free Software, Free Society: Selected Essays of Richard M. Stallman.* Boston, MA: Free Software Foundation, 2006.

Steinem, Gloria. *My Life on the Road.* New York: Random House, 2015.

Stonbely, Sarah. *Comparing Models of Collaborative Journalism.* Montclair, NJ: Center for Cooperative Media, 2018.

Stoney, *George.* "George Stoney Visits American University." Ed. Michael T. Miller and Maura Ugarte, Center for Social Media & Impact, n.d., https://cmsimpact.org/resource/george-stoney-visits-american-univeristy/.

Syms, Martine. "The Mundane Afrofuturist Manifesto." *Rhizome*, December 17, 2013. https://rhizome.org/editorial/2013/dec/17/mundane-afrofuturist-manifesto.

Szebeko, Deborah, and Lauren Tan. "Co-Designing for Society." *Australasian Medical Journal AMJ* 3, no. 9 (2010): 580–590.

Tahir Hemphill Studio. https://www.tahirhemphill.com.

Taylor, Astra. "The Automation Charade." *Logic: A Magazine about Technology*, August 1, 2018, https://logicmag.io/05-the-automation-charade.

Teachout, Zephyr, and Pat Garofalo. "Cuomo Is Letting Billionaires Plan New York's Future: It Doesn't Have to Be This Way." *The Guardian*, May 14, 2020, https://www.theguardian.com /commentisfree/2020/may/14/andrew-cuomo-bill-gates-eric-schmidt-coronavirus.

Tech as Art: Supporting Artists Who Use Technology as a Creative Medium. National Endowment of the Arts, 2021, https://www.arts.gov/sites/default/files/Tech-as-Art-081821.pdf.

"Technology as a Creative Medium: Supporting Artists Who Use Digital Tools to Illuminate Our World." Washington, DC: National Endowment for the Arts, June 2021.

Tegmark, Max. *Life 3.0: Being Human in the Age of Artificial Intelligence*. New York: Knopf, 2017.

Terranova, Tiziana. *Network Culture: Politics for the Information Age*. London: Pluto, 2004.

The Mentor. "Hacker's Manifesto." *Phrack Magazine* 7 (September 25, 1986). http://phrack.org/issues /7/3.html.

Thomas, Stephen Richard. "Hope: Towards an Ethical Framework of Collaborative Practice in Documentary Filmmaking." Master's thesis, University of Melbourne, 2010. https://minerva -access.unimelb.edu.au/bitstream/handle/11343/36279/272089_ST_MA_HOPE.pdf?sequence=1.

Thomas, Verena. "Indigenising Research through A/r/tography: A Case Study of a Collaborative Filmmaking Project in Papua New Guinea." *UNESCO Observatory E-Journal Multi-Disciplinary Research in the Arts* 3, no. 1 (2013): 1–14. http://eprints.qut.edu.au/99667.

Thompson, Nato. *Living as Form: Socially Engaged Art from 1991–2011*. Cambridge, MA: MIT Press, 2012.

Truth and Reconciliation Commission of Canada. *Honouring the Truth, Reconciling for the Future: Summary of the Final Report of the Truth and Reconciliation Commission of Canada*. Winnipeg, Canada: National Center for Truth and Reconciliation, 2015. https://ehprnh2mwo3.exactdn.com /wp-content/uploads/2021/01/Executive_Summary_English_Web.pdf.

Tsing, Anna. *The Mushroom at the End of the World: On the Possibility of Life in Capitalist Ruins*. Princeton, NJ: Princeton University Press, 2017.

Tufekci, Zeynep. *Twitter and Tear Gas: The Power of Fragility of Networked Protest*. New Haven, CT: Yale University Press, 2017.

Turkle, Sherry. "There Will Never Be an Age of Artificial Intimacy." *New York Times*, August 11, 2018, https://www.nytimes.com/2018/08/11/opinion/there-will-never-be-an-age-of-artificial-intimacy .html.

Turkle, Sherry. *The Second Self: Computers and the Human Spirit*. Cambridge, MA: MIT Press, 2005.

Turner, Fred. *From Counterculture to Cyberculture: Stewart Brand, the Whole Earth Network and the Rise of Digital Utopianism*. Chicago: University of Chicago Press, 2006.

Underwood, Corinna. "Automated Journalism—AI Applications at New York Times, Reuters, and Other Media Giants." *Emerj*, November 17, 2019, https://emerj.com/ai-sector-overviews /automated-journalism-applications.

Uricchio, William. "The Algorithmic Turn: Photosynth, Augmented Reality, and the Changing Implication of the Image." *Visual Studies* 26, no. 1 (2011): 25–35.

Uricchio, William. "Data, Culture and the Ambivalence of Algorithms." In *The Datafied Society: Studying Culture through Data*, ed. Mirko Tobias Schäfer and Karin van Es, 125–138. Amsterdam: University of Amsterdam Press, 2017.

Uricchio, William. "Recursive Media." *Questions de communication* (forthcoming, 2022).

Uricchio, William, and Sarah Wolozin. *Mapping the Intersections of Two Cultures: Interactive Documentary and Digital Journalism*. Cambridge, MA: MIT Open Documentary Lab, 2016. http://opendoclab.mit.edu/interactivejournalism/Mapping_the_Intersection_of_Two_Cultures_Interactive_Documentary_and_Digital_Journalism.pdf.

"Values." *Eyebeam,* n.d., https://www.eyebeam.org.

"Values Statements." New Public, n.d., https://newpublic.org/purpose/values-statements.

Varnelis, Kazys. *Networked Publics*. Cambridge, MA: MIT Press, 2012.

Vertov, Dziga. "Kinopravda and Radiopravda." In *Kino Eye: The Writings of Dziga Vertov*, ed. Annette Michelson, 52–56. Berkeley: University of California Press, 1984.

Vertov, Dziga. "We: Variant of a Manifesto." *Kino-Fot* (August 1922): 5.

Vincent, James. "Twitter Taught Microsoft's AI Chatbox to be a Racist Asshole in Less Than a Day." *The Verge*, March 24, 2016, https://www.theverge.com/2016/3/24/11297050/tay-microsoft-chatbot-racist.

Viner, Katharine. "The Rise of the Reader: Journalism in the Age of the Open Web." *The Guardian*, October 9, 2013, https://www.theguardian.com/commentisfree/2013/oct/09/the-rise-of-the-reader-katharine-viner-an-smith-lecture.

Vitello, Paul. "George C. Stoney, Documentary Filmmaker, Dies at 96." *New York Times*, July 15, 2012, https://www.nytimes.com/2012/07/15/arts/television/george-c-stoney-documentarian-dies-at-96.html.

"VoIP Drupal." Drupal, November 15, 2018, https://www.drupal.org/project/voipdrupal.

"Watershed Sandbox: A How-To Guide." Watershed Sandbox, n.d., https://www.watershed.co.uk/sites/default/files/publications/2018-10-08/watershedsandbox_ahowtoguide.pdf.

Waugh, Thomas, Michael Brendan Baker, and Ezra Winton, eds. *Challenge for Change: Activist Documentary at the National Film Board of Canada*. Montreal: McGill-Queen's University Press, 2010.

Webb, Jeff. "Remembering Colin Low and the Fogo Process." *Acadiensis: Journal of the History of Atlantis Region* (blog), March 29, 2016, https://acadiensis.wordpress.com/2016/03/29/remembering-colin-low-and-the-fogo-process.

Weight, Jenny. *The Participatory Documentary Cookbook: Community Documentary Using Social Media*. Melbourne, Australia: RMIT University, 2012.

Williams, Gisella. "Are Artists the New Interpreters of Scientific Innovation?" *New York Times Style Magazine*, September 12, 2017, https://www.nytimes.com/2017/09/12/t-magazine/art/artist-residency-science.html.

Williams, T. F. D., and M. Oakescott, *Cambridge Journal* 7 (1953): 314.

Willis, Deborah, and Natasha L. Logan. *Question Bridge: Black Males in America*. New York: Aperture Foundation and Campaign for Black Male Achievement, 2015.

Winger-Bearskin, Amelia. "Before Everyone Was Talking about Decentralization, Decentralization Was Talking to Everyone." *Immerse*, July 2, 2018, https://immerse.news/decentralized-storytelling-d8450490b3ee.

Winston, Brian, ed. *The Documentary Film Book*. London: Palgrave Macmillan, 2013.

Wissot, Lauren. "Whose Story? Five Doc-Makers on (Avoiding) Extractive Filmmaking." International Documentary Association, September 28, 2017, https://www.documentary.org/feature/whose-story-five-doc-makers-avoiding-extractive-filmmaking.

Worth, Sol, John Adair, and Richard Chalfen. *Through Navajo Eyes*. Albuquerque: University of New Mexico Press, 1972.

Zavala, Diego, and Salvador Leetoy. "Documental colaborativo como herramienta de agencia cultural: Salta, un caso de estudio." *Revista Científica de Información y Comunicación* 13 (2016): 235–261.

Zeidler, Maryse. "Inuit Film Producers Partner with Haida for First-of-Its-Kind Feature." *CBC News*, January 15, 2017, https://www.cbc.ca/news/canada/british-columbia/the-fast-runner-haida-film-1.3935746.

Zimmerman, Arely M. "Documenting DREAMs: New Media, Undocumented Youth and the Immigrant Rights Movement." Connected Learning Alliance, 2012, https://dmlcentral.net/wp-content/uploads/files/documenting_dreams_-_working_paper-mapp_-_june_6_20121.pdf.

Zimmermann, Patricia R. *Documentary across Platforms: Reverse Engineering Media, Place, and Politics*. Bloomington: Indiana University Press, 2017.

Zimmerman, Patricia R., and Helen De Michiel. *Open Space New Media Documentary: A Toolkit for Theory and Practice*. New York: Routledge, 2018.

Zirión, Antonio. "Miradas cómplices: Cine etnográfico, estrategias colaborativas y antropología visual aplicada." *Iztapalapa, Revista de Ciencias Sociales y Humanidades* 78 (2015): 45–70.

Zoettl, Peter Anton. "Creating Images, Creating Identity: Participatory Filmmaking as an Anthropological Praxis." *Journal des Anthropologues* (2012): 53–78. http://jda.revues.org/5093.

Zuboff, Shoshana. *The Age of Surveillance Capitalism: The Fight for a Human Future at the New Frontier of Power*. New York: PublicAffairs, 2019.

Zuckerman, Ethan. *Rewire: Digital Cosmopolitans in the Era of Connection*. New York: Norton, 2013.

Zylinska, Joanna. *Minimal Ethics for the Anthropocene*. Ann Arbor: Open Humanities Press, Michigan Publishing, 2014.

BIOGRAPHIES OF AUTHORS AND COAUTHORS

Juanita Anderson is a veteran producer and documentary filmmaker who proudly hails from Detroit, Michigan. She currently heads the Media Arts and Studies program at Wayne State University. She is a coauthor of the section "Co-Creation and Equity, Five Media Makers of Color Speak Out."

Maria Agui Carter is an Indigenous Latinx/Chinese immigrant who grew up undocumented in New York City and graduated from Harvard University. She is an award-winning filmmaker (Iguanafilms.com), teaches as an assistant professor at Emerson College, and serves on the Diversity Coalition of the Writers Guild of America. She is a coauthor of the section "Co-Creation and Equity, Five Media Makers of Color Speak Out."

Katerina Cizek is a Peabody- and two-time Emmy-winning documentarian and the artistic director of the Co-Creation Studio at the MIT Open Documentary Lab. For over a decade at the National Film Board of Canada, she helped redefine the organization as a digital storytelling hub through her long-form, co-creative documentary projects *Filmmaker in Residence* and *HIGHRISE*. She is the author (with Uricchio) of this book and coprincipal investigator of the field study.

The **Detroit Narrative Agency (DNA)** incubates quality and compelling stories that shift the dominant narratives about Detroit toward liberation and justice, in collaboration with an ecosystem of community members, storytellers, media makers, and organizers. Detroit Narrative Agency coresearchers and co-designers coauthored the sections "If You're Not at the Table, You're on the Menu," "DNA Photo Essay," and "Community Benefits Agreements."

Thomas Allen Harris is an interdisciplinary artist who uses media, photography, and performance to explore family and identity in a participatory model of filmmaking he has been pioneering since 1990. He is the host, director, and executive producer of *Family Pictures USA*, a national PBS series that examines America through the lens of the family photo album. He is on faculty at

Yale University and coauthored the section "Co-Creation and Equity, Five Media Makers of Color Speak Out."

Maori Karmael Holmes, a filmmaker, writer, and curator, is the founder and artistic director of the BlackStar Film Festival. Her works as a filmmaker have screened internationally. She is a coauthor of the section "Co-Creation and Equity, Five Media Makers of Color Speak Out."

Richard Lachman is a professor who directs the Zone Learning network of incubators at Ryerson University. He helped organize the breakout sessions during the Collective Wisdom symposium and drafted contributed summaries for the field study.

Louis Massiah is a documentary filmmaker and founder and director of the Scribe Video Center in Philadelphia. He contributed the section "Origins of Scribe Video Center."

Cara Mertes is the former director of Just Films at the Ford Foundation and contributed "Appropriate Models for Co-Creation."

Sara Rafsky is a writer, researcher, and film producer who works at the intersection of journalism, press freedom, human rights, and documentary film. She contributed spotlights on collaborative journalism, the group Hyphen-Labs, and the Quipu and Question Bridge projects.

Michèle Stephenson is a Brooklyn-based media maker, author, and artist who pulls from her Haitian and Panamanian roots to tell complex intimate stories by, for, and about communities of color. She and her partner, Joe Brewster, cofounded the multiple award-winning media production company, The Rada Film Group. She is a coauthor of the section "Co-Creation and Equity, Five Media-Makers of Color Speak Out."

William Uricchio is a professor of comparative media studies at MIT and professor emeritus at Utrecht University (Netherlands). He is founder and principal investigator of the Open Documentary Lab. He is principal investigator of the field study and author (with Cizek) of this book.

Amelia Winger-Bearskin is an artist, creative director, and organizer who develops cultural communities at the intersection of art, technology, and education. She is now a Banks family preeminence endowed chair and associate professor of artificial intelligence and the arts at the Digital Worlds Institute at the University of Florida. She is Haudenosaunee (Iroquois) of the Seneca-Cayuga Nation of Oklahoma, Deer Clan. She contributed the section "Decentralized Storytelling."

Sarah Wolozin is the founding director of the MIT Open Documentary Lab and the cofounder of the Co-Creation Studio. She is a coprincipal investigator of the field study and contributed spotlights on Stephanie Dinkins and Sougwen Chung.

INDEX

Page numbers followed by f indicate figures.